Mathematical Physics Studies

The series publishes research monographs that address physics problems within a rigorous mathematical framework, highlighting recent developments at the intersection of mathematics and theoretical physics. Research areas covered include dynamical systems, equilibrium and nonequilibrium statistical mechanics, general relativity, classical and integrable systems, many-body quantum systems and condensed matter physics, partial differential equations, probability and random structures, particle physics, quantum field theory, quantum information, quantum mechanics and spectral theory, and string theory and quantum gravity. Each monograph delves into these areas, emphasizing their mathematical foundations and applications in modern physics.

Timothy J. Burchell • Thomas J. Bridges

Transverse Instability of Solitary Waves

Multisymplectic Dirac Operators and the Evans Function

Springer

Timothy J. Burchell
Department of Mathematics
University of Surrey
Guildford, Surrey, UK

Thomas J. Bridges
Department of Mathematics
University of Surrey
Guildford, Surrey, UK

ISSN 0921-3767 ISSN 2352-3905 (electronic)
Mathematical Physics Studies
ISBN 978-981-95-2045-9 ISBN 978-981-95-2046-6 (eBook)
https://doi.org/10.1007/978-981-95-2046-6

Mathematics Subject Classification: 35Q35, 74J35, 37K45, 53D12

This work was supported by EPSRC Doctoral Training Partnership (EP/M508160/1 and EP/R513350/1) and EPSRC and Newton Institute (EP/R014604/1).

This Springer imprint is published by the registered company Springer Nature Singapore Pte Ltd.
The registered company address is: 152 Beach Road, #21-01/04 Gateway East, Singapore 189721, Singapore

Preface

A theoretical framework for the spectral problem, associated with transverse instability of line solitary wave solutions, uniformly travelling in a horizontal plane and embedded in two space dimensions, in the context of Hamiltonian partial differential equations (PDEs), is developed. Rather than individual PDEs, we work with an abstract class of Hamiltonian PDEs, in two space dimensions and time, that are based on multisymplectic Dirac operators, and their generalisation. This class of PDEs models a wide range of nonlinear wave equations. The advantage is that there is a distinct symplectic structure associated with each space dimension and time. These symplectic structures feed a range of geometric properties into both the existence (variational principle, Maslov index, transversality condition) and linear stability problem (multisymplectic partition of the Evans function). The spectral problem associated with the linearisation about a solitary wave is formulated as a dynamical system, with three symplectic structures feeding into the analysis. A two-parameter (spectral parameter and transverse wavenumber) Evans function is constructed from this dynamical system. The structure enables new results to be proved about the Evans function and the linear transverse instability of solitary waves. The main result is an abstract derivative formula for the two-parameter Evans function, for small values of the stability exponent and transverse wavenumber. To illustrate the theory, a class of vector-valued nonlinear wave equations in $2 + 1$ is introduced, which is multisymplectic and designed to have explicit solitary wave solutions. In this example, the stable and unstable subspaces in the Evans function construction are each of dimension four, and can be explicitly constructed. The example is used to illustrate the geometric instability condition, and the inner workings of the theory, and it is studied in detail.

Guildford, UK

Timothy J. Burchell
Thomas J. Bridges

Acknowledgement The first author acknowledges support from an EPSRC Doctoral Training Partnership, grant numbers EP/M508160/1 and EP/R513350/1. The second author thanks the Isaac Newton Institute for Mathematical Sciences, Cambridge, for support and hospitality during the programme Dispersive Hydrodynamics where part of the work on this monograph was undertaken in Autumn 2022. The Newton Institute programme was supported by EPSRC grant number EP/R014604/1.

Contents

1 **Introduction** ... 1

2 **Literature Review** ... 15
2.1 Evans Function Applied to Transverse Instability ... 20
2.2 Nonlinear Stability and Instability ... 23
2.3 Transverse Instability for NLS-Type Equations ... 24
2.4 The Maslov Index of Solitary Waves ... 26
2.5 Transverse Instability for Water Waves ... 27

3 **Multisymplectic Wave Equations and Dirac Operators** ... 31
3.1 Classical Wave Equation as a Dirac Equation ... 32
3.2 The Total Exterior Algebra (TEA) Bundle ... 36
3.3 Variation of Θ and Dirac Operators ... 37
3.4 Clifford Algebra Structure ... 39
3.5 Example of $\mathbf{J}_\partial$ on $M = \mathbb{R}^{1,1}$ ... 40
3.5.1 Example: Nonlinear Wave Equation on $\mathbb{R}^{1,1}$ ... 42
3.6 Construction of $\mathbf{J}_\partial$ on $M = \mathbb{R}^{1,2}$ ... 43
3.6.1 Hermitian Representation in the $2 + 1$ Case ... 45
3.7 Example: Vector-Valued Wave Equation in $\mathbb{R}^{2,1}$ ... 45
3.8 Example: Multisymplectifying the KP Equation ... 47
3.9 Example: Generalised Klein–Gordon for Water Waves ... 49

4 **Solitary Wave Solutions and Their Properties** ... 53
4.1 The Form of the Solitary Wave Solution ... 54
4.2 Reversibility in x and y ... 56
4.2.1 Reversible Solitary Waves ... 57
4.3 Symmetry of $I_1(c, \ell)$ and $I_2(c, \ell)$ ... 57
4.4 Example: the Kadomtsev–Petviashvili Equation ... 59

5 **Linearisation About Solitary Waves** ... 63
5.1 The Eigenfunctions in the Linearisation ... 64

5.2 Transversality and the Lazutkin–Treschev Invariant 67
5.2.1 Orientation 69
5.3 Asymptotics of $\widehat{Z}_\xi$ and the Factor χ 70

6 Spectral Stability and the Evans Function 73
6.1 Constructing the Evans Function 75
6.2 Matrix Definition of the Evans Function 80
6.3 Small (λ, ν) Asymptotics and Derivatives 85

7 Derivatives of the Evans Function 87

8 Summary of Hypotheses Used 99

9 Example: Nonlinear Wave Equation in $2 + 1$ 103
9.1 Linearisation About the Solitary Wave 106
9.2 Eigenvectors and the Factor χ 108
9.3 Momentum Derivatives 111
9.4 The Lazutkin–Treschev Invariant 113
9.5 The Remaining Hypotheses 118
9.6 The Adjoint Equation 122
9.7 Summary 131

10 Concluding Remarks 133

A Multisymplectifying the KP Equation 137

B Spectral Problem with sech^2 Potential 141

C Choosing the Coupling Terms for the 2+1 Example 143
C.1 Applying to the Linearisation 144
C.2 Placing Constraints on the β's 146

D Parity of the Maslov Index of Solitary Waves 149

References 157

Index 163

Nomenclature

$\mathbf{B}$	Hessian of $S(Z)$ evaluated on solitary wave (5.3)
$\mathbf{B}^\infty$	Limit of $\mathbf{B}$ as $\xi \to \pm\infty$ (5.8)
$\bigstar$	Hodge star operator introduced in (3.15)
$\Box$	d'Alembertian operator (3.3)
$\mathbf{w}_j$	Dual Jost solutions introduced in (6.17)
χ	χ along with $\chi^\pm$ are normalisation constants (5.31)
σ^{cont}	Continuous spectrum of the system at infinity (6.9)
δ	Co-differential operator introduced in (3.15)
$\mathcal{A}(M)$	Total exterior algebra bundle (3.29)
$\mathcal{A}^k(M)$	Differential k-forms on M (3.6)
ℓ	The skewness parameter of line solitary waves (1.3)
$\mathbf{J}(c,\ell)$	Primary symplectic operator, defined in (1.27)
$\mathbf{J}_\partial$	Multisymplectic Dirac operator (3.24)
λ	Stability exponent (1.32)
$[\![\cdot,\cdot]\!]_k$	Inner product on k-forms (3.7)
$\mathcal{R}$	Matrix associated with the metric on flat space-time (3.5)
$\mathcal{R}^{(k)}$	Induced metric on TEA vector space (3.47)
$\mathscr{C}\ell_{q,p}$	Clifford algebra on $\mathbb{R}^{q,p}$ in §3.4
μ_j	Eigenvalues of the system at infinity (5.9)
ν	Transverse wavenumber of the perturbation (1.32)
$\Omega_{(c,\ell)}$	Symplectic form $\langle \mathbf{J}(c,\ell)\cdot,\cdot\rangle$ defined in (1.28)
Π	Lazutkin–Treschev invariant (5.21)
ψ	The Souriau mapping (D.3)
$\sigma(\mathbf{A}^\infty)$	Spectrum of the system at infinity (6.7)
Maslov	Maslov index of pairs of Lagrangian planes, in Appendix D
$\Theta(Z)$	Canonical form on the TEA bundle (3.32)
ϖ	Volume form on space-time (3.4)
$\widehat{Z}$	Basic line solitary wave (1.29)
$\widetilde{U(n)}$	Universal cover of the Lagrangian Grassmannian (D.1)
ξ	Moving frame coordinate, $\xi = x + ct + \ell y$ introduced in (1.3)

Ξ_j	Adjoint eigenvectors defined in (5.11)
ζ_j	Eigenvectors of the system at infinity (5.9)
c	Moving frame speed for basic line solitary wave (1.3)
$D(\lambda, \nu)$	Matrix representation of the Evans function in (6.30)
$D(\lambda, \nu)\varphi$	Evans function on $\bigwedge^8(\mathbb{C}^8)$ introduced in (6.14)
E^s	ξ-dependent stable subspace (5.15)
E^u	ξ-dependent unstable subspace (5.15)
$H(Z)$	Hamiltonian for time-direction flow, defined in (1.31)
$I_1(c, \ell)$	I_1 evaluated on the solitary wave (1.16)
$I_1(Z)$	x-direction momentum (1.30)
$I_2(c, \ell)$	I_2 evaluated on the solitary wave (1.16)
$I_2(Z)$	y-direction momentum flux (1.30)
L	Lagrangian density for multisymplectic equation (4.4)
$S(Z)$	Generalised Hamiltonian function (1.26)
V	Burchell's potential for the example in (9.2)
$W(n, \mathbb{C})$	Space of symmetric unitary matrices (D.2)
Z	Coordinates of multisymplectic phase space
$\mathbf{A}$	ξ, λ, ν)-dependent matrix in the Evans ODE (1.35)
$\mathbf{A}^{\pm}$	Limit of $\mathbf{A}$ as $\xi \to \pm\infty$ (1.36)
$\mathbf{a}^{\pm}$	1st tangent vector (Jost solution) to the stable/unstable $(+/-)$ manifolds (5.15)
$\mathbf{b}^{\pm}$	2nd tangent vector (Jost solution) to the stable/unstable $(+/-)$ manifolds (5.15)
$\mathbf{c}^{\pm}$	3rd tangent vector (Jost solution) to the stable/unstable $(+/-)$ manifolds (5.15)
$\mathbf{K}$	Symplectic operator in the x-direction (1.26)
$\mathbf{M}$	Symplectic operator in the time direction (1.26)
$\mathbf{N}$	Symplectic operator in the y-direction (1.26)
$\mathbf{R}_x$	Reversor in the x-direction (4.12)
$\mathbf{R}_y$	Reversor in the y-direction (4.13)
$\mathbf{u}_j$	Solutions of the Evans function ODE (Jost solutions) in (6.12)

Chapter 1
Introduction

Given an oblique line solitary wave, uniform along the crest, travelling in a fixed direction at constant speed, in the (x, y)-plane, a transverse linear instability has a component of its eigenfunction which is transverse to the direction of travel. A typical oblique line solitary wave is shown in Fig. 1.1. Transverse instability can also arise in higher dimensions (an n-space dimension and time solution unstable in $n + 1 + m$, $m > 0$ dimensions), but it is the one space dimension and time $(1 + 1)$ to two space dimensions and time $(2 + 1)$ case that will be of interest here. The spectral problem, associated with the linearisation about a line solitary wave, will be the principal focus. The context is Hamiltonian partial differential equations (PDEs).

The concept of transverse instability of solitary waves appears to have been first discussed in the paper of Kadomtsev and Petviashvili [85], which also introduced the now well-known KP equation,

$$\left(u_t + uu_x + u_{xxx}\right)_x + \sigma u_{yy} = 0\,, \tag{1.1}$$

where $\sigma = \pm 1$. The Korteweg-de Vries (KdV) solitary wave is an exact solution, and linearisation of (1.1) leads to a spectral problem that can be solved many ways (see literature review in Chap. 2).

One of the reasons that transverse instabilities are interesting is that they provide a mechanism for a one-dimensional solution to transition into a solution in higher space dimensions. An example for illustration, that has been well studied, is the Zakharov–Kuznetsov (ZK) equation

$$u_t + uu_x + u_{xxx} + u_{yyx} = 0\,, \tag{1.2}$$

for the scalar-valued function $u(x, y, t)$ [142]. It is a Hamiltonian PDE, and has an explicit line solitary wave solution, similar to that shown in Fig. 1.1. This solitary wave is transversely unstable (see literature review for references). We use the

T. J. Burchell, T. J. Bridges, *Transverse Instability of Solitary Waves*, Mathematical Physics Studies, https://doi.org/10.1007/978-981-95-2046-6_1

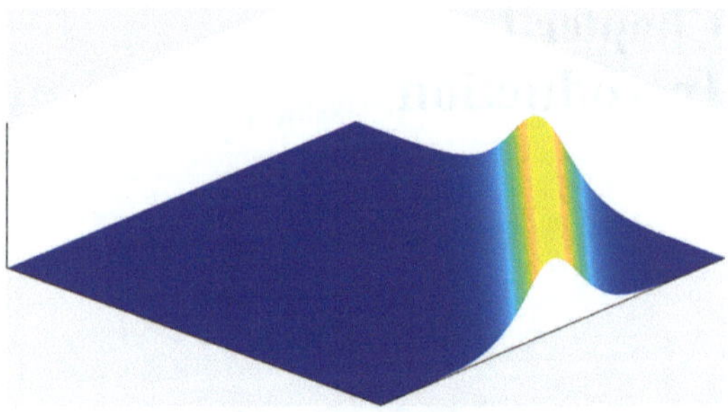

Fig. 1.1 An oblique line solitary wave of amplitude A travelling in the (x, y)-plane

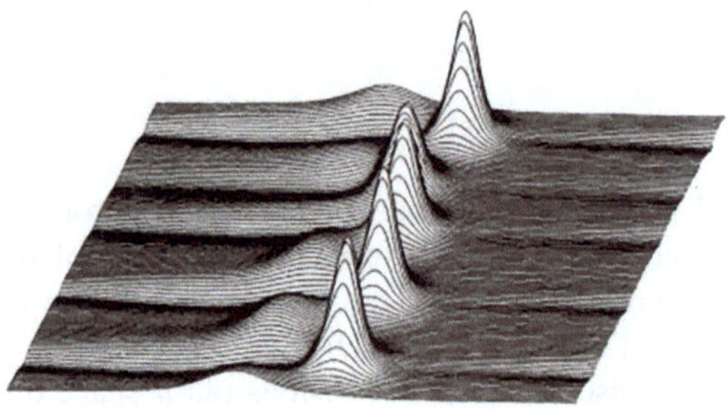

Fig. 1.2 Break-up of a line solitary wave via transverse instability into a field of cylindrical solitons. This is Figure 4 from [81]

ZK example here to illustrate the long-time nonlinear implications of transverse instability. After the linear instability is fully developed, nonlinearity drives the line solitary wave to break-up. Iwasaki et al. [81] showed, via numerical simulations, that the line solitary wave breaks up into a sea of cylindrical solitons which then interact like vortices. An example of this field is shown in Fig. 1.2. Each cylindrical soliton, when well-separated, is an exact solution of the ZK equation [142]. When they are no longer well-separated they dynamically interact creating a fully two-space dimensional complex field of cylindrical vortices. In summary, transverse instability can drive a coherent $1+1$ line solitary wave into a complex incoherent $2+1$ wavefield.

In this monograph, the basic state is a localised solution in the (x, y)-plane, which is uniformly travelling at some speed $c > 0$ along a line in the (x, y)-plane,

$$u(x, y, t) := \widehat{u}(\xi), \quad \xi = x + ct + \ell y, \tag{1.3}$$

where $\ell > 0$ gives the skewness from the x-axis. The function $\widehat{u}(\xi)$ is required to decay exponentially to zero as $\xi \to \pm\infty$. When the governing PDE is isotropic in the (x, y)-plane it is trivial to rotate coordinates so that $\ell = 0$. However, we will not do this because derivatives of the basic state with respect to ℓ will feed into the linear stability problem. After these derivatives are taken, ℓ can be set to zero. In other words, the sensitivity of the solitary wave to rotation in the plane carries information about linear stability (see below in (1.19) how the sensitivity derivatives emerge from a variational principle).

Another example of transverse instability, which is simple enough to study explicitly, and highlights some of the basic features of the linear stability problem, is the following nonlinear wave equation,

$$u_{tt} - u_{xx} - u_{yy} + u - u^2 = 0\,, \tag{1.4}$$

for the scalar-valued function $u(x, y, t)$ in the whole space $(x, y) \in \mathbb{R}^2$. This equation is Hamiltonian with

$$\begin{bmatrix} 0 & -1 \\ 1 & 0 \end{bmatrix} \begin{pmatrix} u \\ v \end{pmatrix}_t = \begin{pmatrix} \delta H/\delta u \\ \delta H/\delta v \end{pmatrix}, \tag{1.5}$$

and

$$H(u, v, u_x, u_y) = \iint_{\mathbb{R}^2} \left[\frac{1}{2} \left(u_x^2 + u_y^2 + u^2 + v^2 \right) - \frac{1}{3} u^3 \right] \mathrm{d}x\mathrm{d}y\,. \tag{1.6}$$

The nonlinear wave equation (1.4) has an explicit solitary-wave solution, which, for the purposes here, can be taken to be aligned with the x-axis,

$$u(x, y, t) = \widehat{u}(x) := 6\kappa^2 \mathrm{sech}^2(\kappa x)\,, \quad \text{with } \kappa = \pm\frac{1}{2}\,. \tag{1.7}$$

Linearise (1.4) about this line solitary wave

$$u_{tt} - u_{xx} - u_{yy} + \big(1 - 2\widehat{u}(x)\big)\, u = 0\,, \tag{1.8}$$

and introduce the spectral ansatz

$$u(x, y, t) = w(x; \lambda, \nu)\mathrm{e}^{\lambda t + \mathrm{i}\nu y}\,, \quad \lambda \in \mathbb{C}\,, \quad \nu \in \mathbb{R}\,. \tag{1.9}$$

The linearised stability problem is then reduced to a spectral problem for $(\lambda, w(x; \lambda, \nu))$, for fixed $\nu \in \mathbb{R}$, where $\lambda \in \mathbb{C}$ is the eigenvalue and $w(x; \lambda, \nu) \in L^2(\mathbb{R})$ is the eigenfunction, satisfying

$$w_{ss} + 12\mathrm{sech}^2(s)w = \Lambda(\lambda, \nu)w\,, \quad s = \kappa x\,, \quad \Lambda = 4(1 + \lambda^2 + \nu^2)\,. \tag{1.10}$$

The basic state $\widehat{u}$ is said to be linearly, spectrally, unstable if there exists, for some $\nu \in \mathbb{R}$, a pair $(\lambda, w(x; \lambda, \nu))$ such that the real part of λ is positive and the function $w \in L^2(\mathbb{R})$. Since λ appears in the square here, instability is confirmed when $\lambda^2 > 0$.

The spectral problem (1.10) can be solved exactly and the point spectrum consists of $\Lambda = 1, 4, 9$, giving three pairs of branches parameterised by ν,

$$\lambda^2 + \nu^2 + \frac{3}{4} = 0\,, \quad \lambda^2 + \nu^2 = 0\,, \quad \lambda^2 = \frac{5}{4} - \nu^2\,.$$

The eigenfunctions can also be explicitly constructed (see Appendix B). The first two branches are stable for all $\nu \in \mathbb{R}$, and the third pair,

$$\lambda = \pm\sqrt{\frac{5}{4} - \nu^2}\,,$$

is unstable for $0 \leq \nu^2 < \frac{5}{4}$. The unstable eigenfunction is an arbitrary multiple of

$$\mathrm{e}^{\Re(\lambda)t}\Big(\Re(w(x;\lambda,\nu))\cos(\nu y) - \Im(w(x:\lambda,\nu))\sin(\nu y)\Big)\,, \tag{1.11}$$

since $\Im(\lambda) = 0$. The instability creates a periodic modulation along the crest of wavelength $2\pi/\nu$. Taking $\widehat{u}(x)$ plus a small multiple of (1.11) as initial data in (1.8) leads to an evolution of the line solitary wave into a fully two-dimensional solution in the (x, y)-plane which is localised in the x-direction and periodic in the y-direction. The steady version of (1.4) is the same as the steady ZK equation and so will have exact cylindrical soliton solutions. Therefore the post-instability dynamics is likely to also have a break-up of the line solitary wave into a sea of dynamically-interacting cylindrical solitons. There are no numerical results for (1.4). However, numerical simulations of transversely unstable line solitary waves for a related nonlinear wave equation (the $2 + 1$ two-way Boussinesq equation) are presented in Blyuss et al. [19]. Wave collapse arises in the long-time dynamics there, as the steady equation does not have exact cylindrical soliton solutions.

Although the example (1.4) gives some idea of the form of the curves of eigenvalues, exact solutions for the eigenvalues and eigenfunctions, for transverse linear instability, are not possible in general, especially when the solitary wave solution is only known abstractly. For example, modifying the nonlinear wave equation (1.4) to have an arbitrary nonlinearity

$$u_{tt} - u_{xx} - u_{yy} + V'(u) = 0\,, \tag{1.12}$$

where $V(u)$ is an arbitrary smooth function, already requires new methods.

In this monograph, we develop a theory for transverse instability for a general abstract class of Hamiltonian PDEs, focusing on the spectral problem, which will be tackled using the Evans function framework. Our main result is to give a sufficient condition for instability, for the abstract class of PDEs, using the geometry of the family of oblique solitary waves.

The existence of a family of solitary waves will be taken as a hypothesis. The interesting aspect of the family is the "geometry of the solitary waves" which is based on the representation of the family of solitary waves by a constrained variational principle in the infinite-dimensional PDE, and the secondary representation of the solitary wave as a homoclinic orbit of a finite-dimensional Hamiltonian system. The homoclinic orbit is transversely constructed (transverse intersection of the stable and unstable manifold) which carries an intersection index called the

Lazutkin–Treschev invariant, as well as a Maslov index, and both of these geometric properties feed into the linear stability problem.

The variational characterisation of a solitary wave in $2+1$ proposed herein is a generalisation of the well-known energy–momentum characterisation of solitary waves, of Hamiltonian PDEs in $1+1$, as critical points of the energy (Hamiltonian) on level sets of the momentum (I). The classical energy–momentum theory in $1+1$ is reviewed using the example (1.4) restricted to one space dimension,

$$u_{tt} - u_{xx} + u - u^2 = 0\,, \tag{1.13}$$

with Hamiltonian

$$H(u,v) = \iint_{\mathbb{R}} \left[\frac{1}{2}\left(u_x^2 + u^2 + v^2\right) - \frac{1}{3}u^3 \right] \mathrm{d}x\,.$$

Travelling waves, of the form $u(x,t) = \widehat{u}(\xi)$ and $v(x,t) = \widehat{v}(\xi)$, with $\xi = x + ct$, when substituted into (1.5) can be interpreted as critical points of the Hamiltonian on level sets of the momentum, with Lagrange necessary condition

$$\delta(H - cI) = 0\,, \tag{1.14}$$

where δ is short-hand notation for variational derivative. For the nonlinear wave equation (1.13) this is

$$\begin{pmatrix} \delta H/\delta u \\ \delta H/\delta v \end{pmatrix} = c \begin{bmatrix} 0 & -1 \\ 1 & 0 \end{bmatrix} \begin{pmatrix} u \\ v \end{pmatrix} = \begin{pmatrix} \delta I/\delta u \\ \delta I/\delta v \end{pmatrix},$$

with momentum function

$$I(u,v) = \int_{\mathbb{R}} v u_\xi \,\mathrm{d}\xi\,. \tag{1.15}$$

The solitary wave solution of (1.14), denoted by $\widehat{u}(\xi, c), \widehat{v}(\xi, c)$, depends freely on c. However, substitution into the constraint,

$$I(\widehat{u}(\xi, c), \widehat{v}(\xi, c)) = \widetilde{I}\,, \tag{1.16}$$

creates a mapping between c values and the values $\widetilde{I}$ of the constraint. This mapping is invertible, assuming smooth solutions, when

$$\frac{dI}{dc} \neq 0\,. \tag{1.17}$$

In addition to being important for non-degeneracy of the critical point problem, this mapping, and the sign of the derivative, enters the linear stability problem for solitary waves. For example, the sign of dI/dc appears in the derivative formula for the Evans function, applied to the longitudinal instability theory for solitary waves [35]. In this monograph, we extend both the variational principle, and the derivative formula to $2+1$.

Firstly, we will extend the energy–momentum method to $2+1$ by including a second constraint set. The second constraint set captures the obliqueness of the solitary wave in $2+1$, and it turns out to be associated with the transverse momentum flux of line solitary waves in $2+1$. For an oblique solitary wave of the form (1.3), there are two key parameters, c and ℓ, and they will appear as Lagrange multipliers. The new variational principle, with two constraint sets, is to find critical points of the Hamiltonian H restricted to level sets of two functionals I_1 and I_2,

$$\delta(H - cI_1 - \ell I_2) = 0\,. \tag{1.18}$$

The functional I_1 is still the longitudinal momentum, and we will find that I_2 can be identified with the transverse momentum flux, and an abstract formula for it will be derived. The properties of I_2 will arise naturally once we transform the Hamiltonian PDE with a single symplectic structure to a Hamiltonian PDE on a multisymplectic structure. In the multisymplectic structure there is a distinct symplectic structure for the y-direction and it generates the transverse momentum flux function I_2. The time-direction symplectic structure, transformed to multisymplectic coordinates, generates the momentum (1.15).

The functionals H, I_1 and I_2, and their evaluation on a family of solitary waves will be denoted by the same symbols to simplify notation. The Jacobian matrix of the latter two functionals

$$\begin{bmatrix} \dfrac{\partial I_1}{\partial c} & \dfrac{\partial I_1}{\partial \ell} \\[2mm] \dfrac{\partial I_2}{\partial c} & \dfrac{\partial I_2}{\partial \ell} \end{bmatrix} \tag{1.19}$$

and its determinant, will feed into the analysis of the linear stability problem (see formula (1.38) below). The significance of the Jacobian matrix (1.19) in classical Lagrange multiplier theory, generalising (1.16) and (1.17) is as follows. The Lagrange necessary condition (1.18) produces solutions that depend explicitly on c, ℓ. Denote these critical points of (1.18) by $\widehat{Z}(\xi, c, \ell)$, where, at this point (c, ℓ) are free parameters in the equation. To link them to the values of the constraint sets, substitute the solutions $\widehat{Z}(\xi, c, \ell)$ into the constraints

$$I_1(\widehat{Z}(\xi, c, \ell)) = \widetilde{I}_1 \quad \text{and} \quad I_2(\widehat{Z}(\xi, c, \ell)) = \widetilde{I}_2\,, \tag{1.20}$$

where $\widetilde{I}_1$ and $\widetilde{I}_2$ are the assigned values of the level sets. The expressions (1.20) provide a mapping between (c, ℓ) and $(\widetilde{I}_1, \widetilde{I}_2)$, and this mapping is invertible when the determinant of the Jacobian matrix in (1.19) is non-zero. This invertibility provides the classical significance of (1.19) in Lagrange multiplier theory. However, we will see that (1.19) appears in a different and surprising way in the transverse stability theory for solitary waves.

Instead of developing a stability theory for solitary waves of specific Hamiltonian PDEs, such as Zakharov–Kuznetsov, or the Kadomtsev–Petviashvili equation, or other specific nonlinear wave equations, we will be introducing an abstract class of nonlinear Hamiltonian PDEs. This class of PDEs is based on multisymplectic Dirac operators [27, 33], but will be generalised further to include all of the above PDEs as well as more general multisymplectic Hamiltonian PDEs.

To illustrate the basic idea of a multisymplectic Hamiltonian PDE, the nonlinear wave equation (1.4) is used. It is the Euler–Lagrange equation associated with the Lagrangian density

$$L(u, u_t, u_x, u_y) = \frac{1}{2}u_t^2 - \frac{1}{2}u_x^2 - \frac{1}{2}u_y^2 - V(u)\,, \tag{1.21}$$

where $V(u)$ can be an arbitrary smooth function, and is $V(u) = \frac{1}{2}u^2 - \frac{1}{3}u^3$ for the case (1.4). The classical Legendre transform (transforming the time direction only) generates a Hamiltonian function whose density is

$$H(u, v, u_x, u_y) = vu_t - L(u, u_t, u_x, u_y)\,, \qquad v := \frac{\partial L}{\partial u_t}\,, \tag{1.22}$$

with u_t eliminated in terms of u, v, u_x, u_y by inverting the second equation in (1.22). The Hamiltonian function is obtained by substituting u_t, and integrating over x, y:

$$H(u, v, u_x, u_y) = \int\int_{\mathbb{R}^2} \left(\frac{1}{2}(v^2 + u_x^2 + u_y^2) + V(u)\right) \mathrm{d}x\mathrm{d}y\,, \tag{1.23}$$

which agrees with (1.6) when $V'(u) = u - u^2$. This recovers the classical Hamiltonian evolution formulation (1.5).

One way to derive a multisymplectic formulation is to use a *total Legendre transform*—that is, in x, y, t; define

$$v := \frac{\partial L}{\partial u_t}\,, \qquad w := \frac{\partial L}{\partial u_x}\,, \qquad p := \frac{\partial L}{\partial u_y}\,.$$

These equations can be inverted to give $v = u_t$, $w = -u_x$, and $p = -u_y$. The new generalised Hamiltonian function is then

$$S = vu_t + wu_x + pu_y - L = \frac{1}{2}v^2 - \frac{1}{2}w^2 - \frac{1}{2}p^2 + V(u)\,. \tag{1.24}$$

The governing equations are

$$\begin{aligned} -v_t - w_x - p_y &= \frac{\delta S}{\delta u}\,, \\ u_t &= \frac{\delta S}{\delta v}\,, \\ u_x &= \frac{\delta S}{\delta w}\,, \\ p_y &= \frac{\delta S}{\delta p}\,. \end{aligned} \tag{1.25}$$

This strategy is due to DeVonder and Weyl (see e.g. Chapter 6 of [17]). It has shortcomings as we will see. We will develop more effective strategies for constructing multisymplectic Hamiltonian PDEs. Firstly, we will use relaxed variational principles, following [49, 51], which generalise the Legendre transform, and even for the scalar wave equation (1.4) provides a tidier and more effective Hamiltonian formulation (see Sect. 3.1 in Chap. 3). Secondly, we will develop multisymplectic structures from the structure of Riemannian and pseudo-Riemannian manifolds, following [27, 33, 34], which circumvents the need for a Lagrangian, and ultimately leads to totally non-degenerate multisymplectic structures, as well as multisymplectic Dirac operators (see Sect. 3.2 in Chap. 3).

In addition to the Hamiltonian function S in (1.24), the Legendre transform generates three symplectic structures, based on exterior differentiation of the one forms $v\mathrm{d}u$, $w\mathrm{d}u$, and $p\mathrm{d}u$, giving

$$\mathrm{d}v \wedge \mathrm{d}u\,, \quad \mathrm{d}w \wedge \mathrm{d}u\,, \quad \mathrm{d}p \wedge \mathrm{d}u\,.$$

It is this collection of symplectic (or pre-symplectic) forms which motivatives the name "multisymplectic structure". The above two forms then generate skew-symmetric matrices which generalise the concept of symplectic operators.

Regardless of how the multisymplectic formulation is obtained, we will always put it in standard form

$$\mathbf{M}Z_t + \mathbf{K}Z_x + \mathbf{N}Z_y = \nabla S(Z)\,, \quad Z \in \mathbb{R}^N\,, \tag{1.26}$$

where $\mathbf{M}$, $\mathbf{K}$ and $\mathbf{N}$ are $N \times N$ skew-symmetric matrices and ∇S is the gradient of smooth function $S : \mathbb{R}^N \to \mathbb{R}$, which acts as a generalised Hamiltonian function. The phase space in this formulation is finite-dimensional (in contrast to Hamiltonian evolution equations), and is taken to be $\mathbb{R}^N$ in (1.26) with N a positive even integer. For the wave equation example (1.25), the phase space is $\mathbb{R}^4$, and the structure matrices are

$$\mathbf{M} = \begin{bmatrix} 0 & -1 & 0 & 0 \\ 1 & 0 & 0 & 0 \\ 0 & 0 & 0 & 0 \\ 0 & 0 & 0 & 0 \end{bmatrix}, \quad \mathbf{K} = \begin{bmatrix} 0 & 0 & -1 & 0 \\ 0 & 0 & 0 & 0 \\ 1 & 0 & 0 & 0 \\ 0 & 0 & 0 & 0 \end{bmatrix}, \quad \mathbf{N} = \begin{bmatrix} 0 & 0 & 0 & -1 \\ 0 & 0 & 0 & 0 \\ 0 & 0 & 0 & 0 \\ 1 & 0 & 0 & 0 \end{bmatrix}.$$

The non-degeneracy of these structure matrices can cause problems in the analysis (for example, with Hypothesis (**H1**) in Chap. 3), hence the need for a new way to construct the multisymplectic structure. The formulation (1.26) is most effective when at least one of the structure matrices is invertible. We will find that the most effective formulation is based on a multisymplectic Dirac operator, where *all the structure matrices are non-degenerate*, thereby generating three independent proper symplectic structures on $\mathbb{R}^N$.

It is the abstract form (1.26) that will form the basis of the analysis in this monograph. We will not need the full power of multisymplectic geometry; in the analysis of solitary waves and their stability the key properties of (1.26) that feed into the analysis are just the skew-symmetry of **M**, **K** and **N**, which generate distinct symplectic structures, and the fact that the right-hand side is a gradient. (Strictly speaking, when a constant skew-symmetric matrix has a non-trivial kernel, it generates a *pre-symplectic* structure, which is then symplectic on a subspace. Since the distinction is not important here, we will just use the term "symplectic structure" associated with each of **M**, **K**, and **N** with this understanding.)

The abstract form (1.26) stands alone as a starting point for the analysis. However, we will take the story of (1.26) to another level by generating multisymplectic Hamiltonian PDEs directly from the representation of space-time by a Riemannian or pseudo-Riemannian manifold M. On the manifold M, there exists an intrinsic form on the total exterior algebra bundle of M. The pullback of this differential form generates a functional whose first variation is an example of the left-hand side of (1.26). Remarkably, in this case the matrices **M**, **K**, and **N** are all non-degenerate and form the basis of a Clifford Algebra (cf. [27, 33, 34]). These are the "multisymplectic Dirac operators" referred to above. This construction generalises the symplectic structure on the cotangent bundle of a Riemannian manifold in classical mechanics. The constructive aspects and relevant properties of these multisymplectic operators are given in Chap. 3.

The interplay between the three symplectic structures generated by **M**, **K**, and **N** in (1.26) feeds into the analysis as follows: the time evolution and the energy–momentum plus momentum flux characterisation of the solitary wave use the t-symplectic structure and the y-symplectic structure, the transversality of the homoclinic orbit representation of the solitary wave is defined using a combination of the x-direction, y-direction, and t-direction symplectic structure (see definition of $\mathbf{J}(c, \ell)$ in (1.27) below), and the y-direction symplectic structure will be linked to the transverse stability exponent, in the linear stability analysis. The Evans function analysis will use all three symplectic structures.

The primary symplectic structure in the analysis is a linear combination of the three matrices on the left-hand side of (1.26). Each of the three matrices is not required to be invertible, but the linear combination

$$\mathbf{J}(c,\ell) = \mathbf{K} + c\mathbf{M} + \ell\mathbf{N}\,, \tag{1.27}$$

is required to be invertible, for $(c,\ell) \in \mathscr{C}$, where $\mathscr{C}$ is the set of (c,ℓ) for which a solitary wave exists. In this case $(\mathbb{R}^N, \Omega)$ is a symplectic vector space with symplectic form

$$\Omega_{(c,\ell)}(U,V) := \langle \mathbf{J}(c,\ell)U, V\rangle\,, \quad \forall\, \mathbf{u},\mathbf{v} \in \mathbb{R}^N\,, \quad (c,\ell) \in \mathscr{C}\,, \tag{1.28}$$

where $\langle\cdot,\cdot\rangle$ is the Euclidean inner product on $\mathbb{R}^N$.

This symplectic structure feeds into the existence of solitary waves via the Euler–Lagrange equation (1.18). A solitary wave solution is of the form

$$Z(x,y,t) := \widehat{Z}(\xi)\,, \quad \xi = x + ct + \ell y\,, \tag{1.29}$$

and it is characterised as a critical point of H restricted to level sets of I_1 and I_2. In the multisymplectic setting, the longitudinal momentum and transverse momentum flux, evaluated on a solitary wave, are

$$I_1(Z) = \int_{\mathbb{R}} \frac{1}{2}\langle \mathbf{M}Z_\xi, Z\rangle\, \mathrm{d}\xi \quad \text{and} \quad I_2(Z) = \int_{\mathbb{R}} \frac{1}{2}\langle \mathbf{N}Z_\xi, Z\rangle\, \mathrm{d}\xi\,. \tag{1.30}$$

The Euler–Lagrange equation (1.18) is

$$\nabla H(\widehat{Z}) = c\nabla I_1(\widehat{Z}) + \ell\nabla I_2(\widehat{Z})\,,$$

which can be written out in the form

$$\mathbf{J}(c,\ell)\widehat{Z}_\xi = \nabla H(\widehat{Z})\,, \quad \text{where } \nabla H = \nabla S - \mathbf{K}\widehat{Z}_\xi\,. \tag{1.31}$$

Since $\widehat{Z} : \mathbb{R} \to \mathbb{R}^N$ and $\mathbf{J}(c,\ell)$ is symplectic, for all $(c,\ell) \in \mathscr{C}$, this equation shows that a solitary wave can be characterised as a homoclinic orbit of a finite-dimensional Hamiltonian system. This characterisation will be useful for several aspects of the analysis; for example, such homoclinic orbits, when transversely constructed, have a symplectic intersection index, and when they decay exponentially as $\xi \to \pm\infty$, they have a Maslov index (see Sect. 2.4 of the literature review).

The Evans function theory will be used to analyse the spectral problem associated with the linearisation about a solitary wave. The novelty here is how the three symplectic structures feed into the analysis of the Evans function. The operator $\mathbf{J}(c,\ell)$ is central and arises as follows. Linearising the class of PDEs (1.26) about a family of solitary waves, and introducing a spectral ansatz of the form (1.9),

$$Z(x, y, t) = W(\xi; \lambda, \nu)\mathrm{e}^{\lambda t + \mathrm{i}\nu y}, \quad \lambda \in \mathbb{C}, \quad \nu \in \mathbb{R}, \tag{1.32}$$

leads, in the multisymplectic setting, to a canonical form for the spectral problem

$$\mathbf{J}(c, \ell)W_\xi = \big[\mathbf{B}(\xi, c, \ell) - \lambda\mathbf{M} - \mathrm{i}\nu\mathbf{N}\big]W, \quad W \in \mathbb{C}^N, \tag{1.33}$$

where $\mathbf{B}$ is a ξ-dependent symmetric operator (the Hessian of $S(Z)$ evaluated on a solitary wave) and $\mathrm{i}\mathbf{N}$ is Hermitian. Inverting $\mathbf{J}$ leads to a dynamical systems form of the linear stability problem,

$$W_\xi = \mathbf{A}(\xi, \lambda, \nu)W, \quad W \in \mathbb{C}^N, \tag{1.34}$$

with

$$\mathbf{A}(\xi, \lambda, \nu) = \mathbf{J}(c, \ell)^{-1}\big[\mathbf{B}(\xi, c, \ell) - \lambda\mathbf{M} - \mathrm{i}\nu\mathbf{N}\big]. \tag{1.35}$$

In (1.34), the dependence on (c, ℓ) is suppressed to lighten notation.

The dynamical systems formulation of the spectral problem (1.34) is in standard form for application of the theory of the Evans function. When the solitary wave solution decays exponentially to zero as $\xi \to \pm\infty$, the coefficient matrix $\mathbf{A}(\xi, \lambda, \nu)$ has the property

$$\mathbf{A}(\xi, \lambda, \nu) \to \mathbf{A}^\pm(\lambda, \nu), \quad \text{as } \xi \to \pm\infty. \tag{1.36}$$

It will be assumed that $\mathbf{A}^\pm(\lambda, \nu)$ is hyperbolic for a chosen open set of λ, ν values. Background and history of the Evans function is given in the literature review in Chap. 2, Sect. 2.1.

The Evans function theory has been developed predominantly in $1 + 1$. The theory has been extended to the linearisation about solitary waves in $2 + 1$ by Blyuss et al. [19], Deconinck et al. [61], Yamazaki [140], and Benzoni-Gavage [11]. Closest to the present work is the last paper as the context is Hamiltonian PDEs; in particular, a class of Hamiltonian PDEs based on the Euler–Korteweg equation. The Hamiltonian structure was used in the existence of the solitary wave in [11], but not used in the construction or analysis of the Evans function. Here, we take the $2 + 1$ Evans function analysis in a new direction, by applying it to an abstract class of Hamiltonian PDEs, rather than to particular PDEs, decomposing the governing equation (1.34) into multisymplectic structures, and using the symplectic structures at every step of the analysis.

With the assumption that λ and ν lie in a subset of $\mathbb{C} \times \mathbb{R}$ where the stable and unstable subspaces of $\mathbf{A}^\pm(\lambda, \nu)$ have dimension $N/2$, the Evans function can be constructed as an intersection form for these subspaces. It is a complex analytic function, denoted by $D(\lambda, \nu)$, whose zeros correspond to branches of eigenvalues $\lambda(\nu)$. Further development of the properties of the Evans function are given in

Chap. 6. One of the main outcomes of this theory is a proof that the Evans function to leading order has the property

$$D(\lambda, \nu) = \frac{1}{2} D_{\lambda\lambda}(0,0)\lambda^2 + D_{\lambda\nu}(0,0)\lambda\nu + \frac{1}{2} D_{\nu\nu}(0,0)\nu^2 + o\big(|\lambda|^2 + |\nu|^2\big)\,, \qquad (1.37)$$

with the following explicit formulas for the Taylor coefficients,

$$\begin{aligned} &D(0,0) = 0\,, \quad D_\lambda(0,0) = 0\,, \quad D_\nu(0,0) = 0\,, \quad D_{\lambda\lambda}(0,0) = \chi\Pi\frac{\partial I_1}{\partial c}\,,\\ &D_{\lambda\nu}(0,0) = \mathtt{i}\chi\Pi\frac{\partial I_1}{\partial \ell} = \mathtt{i}\chi\Pi\frac{\partial I_2}{\partial c} \quad \text{and} \quad D_{\nu\nu}(0,0) = -\chi\Pi\frac{\partial I_2}{\partial \ell}\,, \end{aligned} \qquad (1.38)$$

where Π is the Lazutkin–Treschev invariant (a symplectic transversality index, defined in Chap. 5), χ is a property of the homoclinic orbit representing the solitary wave, also defined in Chap. 5, and $I_1(c, \ell)$, $I_2(c, \ell)$ are defined in (1.30), evaluated on the (c, ℓ)-dependent family of solitary waves. The Lazutkin–Treschev invariant is also connected with the Maslov index, of the homoclinic orbit representation of the solitary wave, by

$$\text{sign}(\Pi) = (-1)^{\text{Maslov}}\,. \qquad (1.39)$$

It is transversality, via Π, that arises naturally in the construction of the derivatives of the Evans function, with respect to λ and ν. The connection of transversality with the Maslov index in (1.39) is surprising and comes from the theory for the Maslov index in Chardard and Bridges [40].

The hypotheses under which the result (1.38) holds are summarised in Chap. 8, and the proof is given in Chap. 7. The advantage of this result is that transverse instability of a solitary wave can be deduced from the solitary wave itself. For example, it follows from the above formula that a sufficient condition for transverse instability of the basic solitary wave is

$$\det\begin{bmatrix} \dfrac{\partial I_1}{\partial c} & \dfrac{\partial I_1}{\partial \ell} \\[2ex] \dfrac{\partial I_2}{\partial c} & \dfrac{\partial I_2}{\partial \ell} \end{bmatrix} > 0\,, \qquad (1.40)$$

independent of the Luzutkin invariant, the parameter χ and the orientation of the Evans function. When the determinant in (1.40) is negative, secondary criteria for instability can be derived along rays in the $(\Re(\lambda), \nu)$ plane.

To illustrate the results (1.38) and (1.40), we give a detailed analysis of an example wave equation

$$\Box\phi + \nabla_\phi V = 0 \quad \Box\mathbf{v} + \nabla_\mathbf{v} V = 0\,, \qquad (1.41)$$

where $\phi(x, y, t)$ is scalar-valued and $\mathbf{v}(x, y, t) \in \mathbb{R}^3$. The operator $\square$ is the d'Alembertian,

$$\square\phi := \phi_{tt} - \phi_{xx} - \phi_{yy} ,$$

and the form of the smooth function $V : \mathbb{R}^{1+3} \to \mathbb{R}$ is chosen so that the Evans function can be computed explicitly (see Appendix C). The details of this example are presented in Chap. 9. There are many other examples to which the theory applies. We show that many of the model equations in water waves (e.g. KP, Boussinesq, generalised Klein–Gordon, etc.) are multisymplectic, and sketch how the theory might apply. In the case of the KP equation, the transverse instability formula can be explicitly applied and this is done in Sect. 4.4.

An outline of the monograph is as follows. We begin in Chap. 2 with a literature review, covering stability of solitary waves of Hamiltonian PDEs in 1 + 1, the transverse instability of solitary waves, and the role of the Evans function in 1 + 1 and its extension to 2 + 1, the Maslov index, and applications in optics and water waves. The abstract class of Hamiltonian PDEs based on multisymplectic structures in general and Dirac operators in particular is introduced in Chap. 3. The properties of the existing solitary waves, and the properties of the linearisation about solitary waves are developed in Chap. 4. The linearisation about an oblique line solitary wave in the class of 2 + 1 Hamiltonian PDEs is carried out in Chap. 5, as well as the reduction to a spectral problem. The derivation of the dynamical systems formulation of the spectral problem (1.33), along with the role of multisymplecticity, and the derivation of the Evans function itself, are given in Chap. 6. The derivative formula (1.38) is introduced and proved in Chap. 7. A summary of hypotheses is given in Chap. 8. The 2 + 1 example mentioned above (1.41) is introduced and analysed in Chap. 9. Concluding remarks including comments on open problems are given in Chap. 10. The appendices give detail on special topics: multisymplectifying the KP equation in Appendix A, detail on the spectral problem (1.10) in Appendix B, construction of the potential for the 2 + 1 wave equation (1.41) in Appendix C, and background on the Maslov index of solitary waves in Appendix D.

Chapter 2
Literature Review

A review of previous work on the transverse instability of solitary waves is given in this chapter with an emphasis on topics that resonate with themes in this monograph. The basic state throughout is a solitary wave that is localised and decays exponentially at infinity. Transverse instability of other types of solitary waves as well as periodic travelling waves, although related and important, will be only briefly mentioned. It is the spectral problem associated with the linearisation that is of main interest. Nonlinear stability is not considered, but some results that build on the linear stability problem are reviewed. The principal applications of the theory familiar to the authors are in optics and in the theory of water waves. The former are reviewed briefly in the nonlinear Schŕodinger equation (NLS) section below, and the latter in the section on water waves. Attention will be restricted to PDEs that have been identified as Hamiltonian, or can be shown to be Hamiltonian (classical or multisymplectic).

The transverse instability of solitary waves appears to have been first discussed in the paper of Kadomtsev and Petviashvili [85], which also introduced the KP equation

$$\left(u_t + uu_x + u_{xxx}\right)_x + \sigma u_{yy} = 0\,, \tag{2.1}$$

where $u(x, y, t)$ is scalar-valued, $\sigma = \pm 1$ (with $\sigma = +1$ called KP-II and $\sigma = -1$ called KP-I). The $1 + 1$ KdV solitary wave, satisfying $u_t + uu_x + u_{xxx} = 0$, is an exact solution of this equation, and the question addressed in [85] is instability with respect to y-dependent perturbations, of the form

$$u(x, y, t) = \hat{u}(\xi) + v(\xi, y, t)\,, \quad \xi = x + ct\,, \quad \|v\| \ll 1\,, \tag{2.2}$$

where $\hat{u}(\xi)$ is the KdV solitary wave. Substituting (2.2) into (2.1) and linearising gives

T. J. Burchell, T. J. Bridges, *Transverse Instability of Solitary Waves*, Mathematical Physics Studies, https://doi.org/10.1007/978-981-95-2046-6_2

$$v_{\xi t} + \big(a(\xi)v\big)_{\xi\xi} + v_{\xi\xi\xi\xi} + \sigma v_{yy} = 0$$

(see e.g. §2.5 of Kuznetsov et al. [98] and Alexander et al. [5] for a derivation). With a spectral and normal-mode ansatz,

$$v(x, y, t) = \mathrm{e}^{\lambda t}\widehat{v}(\xi)\mathrm{e}^{\mathbf{i}\nu y}\,, \quad \nu \in \mathbb{R}\,,$$

the linearised equation reduces to a fourth-order differential eigenvalue problem,

$$\lambda\widehat{v}_\xi + \big(a(\xi)\widehat{v}\big)_{\xi\xi} + \widehat{v}_{\xi\xi\xi\xi} - \nu^2\sigma\widehat{v} = 0\,. \tag{2.3}$$

This eigenvalue problem can be solved explicitly (see e.g. [5]), showing that the KdV solitary wave is transversely unstable (stable) when $\sigma = -1$ ($\sigma = +1$). The transverse instability of line solitary waves of the KP equation has since been studied by a wide range of methods (e.g. Ostrowskii and Shrira [113], Infeld et al. [80], Kuznetsov et al. [98], Infeld and Rowlands [79], Kataoka et al. [91], Bridges [24, 25], Kataoka and Tsutahara [89]). The KP equation is Hamiltonian and multisymplectic (see Sect. 3.8 and Appendix A).

The Zakharov–Kuznetsov equation (1.2) has also been widely studied as a prototype for transverse instability. It has an explicit line solitary wave solution that is transversely unstable (e.g. Zakharov and Kuznetsov [142], Allen and Rowlands [6], Munro and Parkes [111], Bridges [24], Rousset and Tzvetkov [126], and Yamazaki [140]). The ZK equation is a useful example to illustrate the long-time fully nonlinear implications of transverse instability, even though this latter development has only been shown numerically. After the linear instability is fully developed, nonlinearity drives the line solitary wave to break-up. Numerical simulations show that the line solitary wave breaks up into a sea of cylindrical solitons, as shown in Fig. 1.2, which then interact like vortices (e.g. Iwasaki et al. [81], Sipcic and Benney [132], Bridges and Reich [32], and Klein et al. [96]). Each cylindrical soliton, when well-separated, is an exact solution of the $2+1$ ZK equation [142]. When they are no longer well-separated they dynamically interact creating a fully two-space dimensional complex sea of cylindrical vortices. The upshot is that a coherent $1+1$ line solitary wave can give rise to a complex incoherent $2+1$ wavefield. The ZK equation is a Hamiltonian PDE and has a multisymplectic formulation [24].

The fifth-order KP equation (KP5) generalises (2.1) to include higher-order dispersion

$$\big(u_t + uu_x + \alpha\eta_{xxx} + u_{xxxxx}\big)_x + \sigma u_{yy} = 0\,, \tag{2.4}$$

for scalar-valued $u(x, y, t)$ with α a real parameter and $\sigma = \pm 1$. The $1+1$ version of this equation is the fifth-order KdV equation, also called the Kawahara equation,

$$u_t + uu_x + \alpha u_{xxx} + u_{xxxxx} = 0\,. \tag{2.5}$$

This equation has classical solitary wave solutions which are localised and decay exponentially to zero at infinity, but it also has a much wider range of solitary waves, including multipulse solitary waves, and generalised solitary waves (solitary waves that are biasymptotic to small but finite-amplitude periodic waves). Haragus and Wahlen [75] consider the transverse instability of generalised solitary waves of KP5 in the form

$$u_{xt} = \left(u_{xxxx} + u_{xx} + \frac{1}{2}u^2\right)_{xx} + u_{yy}\,.$$

They prove that the generalised solitary waves are transversely unstable. Recent results of Sprenger et al. [134] show that KdV5 (2.5) has even more exotic generalised solitary waves where the periodic states at plus and minus infinity are different. Transverse instability, or even longitudinal instability,[1] of states of the type in [75] or [134] are outside of the scope of this monograph but are an intriguing direction for future research.

Cho [48] considers a similar KP5 with slightly different coefficient values,

$$u_{xt} + \left(u_{xxxx} + 2u_{xx} + 3u^2\right)_{xx} + u_{yy} = 0\,, \tag{2.6}$$

He has extensive results using analytical and numerical methods, for both stability of line solitary waves as well as stability results for fully 3D localised solutions. Of interest here is Section IV of [48], where transverse instability of line solitary waves is discussed. He introduces a perturbation as in (2.2) and linearises giving

$$v_{xxxxxx} + 2v_{xxxx} + 6(\hat{u}v)_{xx} + cv_{xx} - \nu^2 v + \lambda v_x = 0\,.$$

Using a long-wave expansion ($|\nu| \ll 1$) he shows that there is transverse instability whenever

$$\frac{\partial}{\partial c}\int_{-\infty}^{\infty} \hat{u}^2\,\mathrm{d}\xi < 0\,,$$

Since the $L^2(\mathbb{R})$ norm is the momentum of KdV5, this is the longitudinal stability condition for KdV5 solitary waves, when the eigenvalue condition of the Hessian is satisfied (e.g. Levandosky [100, 101], Bridges and Derks [30]). Cho uses this property to conclude that the depression and elevation solitary waves, which decay exponentially to zero at infinity, are longitudinally stable but transversely unstable. The theory is accompanied by intriguing numerical simulations. He shows that the $1+1$ *depression* solitary wave, which propagates with speed $c = -2$ in the x-direction, transforms into two 3D depression co-propagating solitary waves. On the

[1] In longitudinal stability analysis, the perturbation travels in the same direction as the solitary wave.

other hand, a $1+1$ *elevation* solitary wave, taken as initial data, also propagating with speed $c = -2$ in the x-direction, transforms into two 3D depression co-propagating solitary waves, as well as two 3D depression waves propagating in oblique directions.

In Borluk et al. [21] transverse instability of a generalised KP with fractional-order dispersion is studied,

$$\left(u_t + uu_x - D^{\alpha} u_x\right)_x + \sigma u_{yy} = 0\,,$$

for scalar-valued $u(x, y, t)$, where $\frac{1}{3} < \alpha \le 2$ and $\sigma = \pm 1$. The linear operator D^{α} denotes the Riesz potential of order α in the x-direction only. They prove that the $1+1$ classical solitary wave solution is linearly transversely unstable.

Transverse stability for smooth solitary waves of a 2+1 Camassa–Holm equation is proved by Geyer et al. [70]. Their equation is

$$(u_t - u_{txx} + 3uu_x - 2u_x u_{xx} - uu_{xxx})_x + u_{yy} = 0\,.$$

They point out that the theory for transverse instability is similar to that for KP-II and they prove that the smooth $1+1$ solitary waves are linearly stable.

Transverse instability for the $(2+1)$-two-way Boussinesq equation

$$u_{tt} + \left(f(u) + \alpha u_{xx}\right)_{xx} + \sigma u_{yy} = 0\,, \tag{2.7}$$

with $\alpha, \sigma = \pm 1$ and $f(u)$ an arbitrary smooth function, with the canonical form $f(u) = D(u^2 - u)$ (with $D = \pm 1$), has been studied analytically and numerically in Blyuss et al. [19], and numerically in Hu et al. [78]. In [19] the basic solitary wave solution is taken to be

$$u(\theta; c, \ell) = A(c, \ell)\,\mathrm{sech}^2\big(B(c, \ell)\theta\big)\,, \tag{2.8}$$

with $\theta = x + ct + \ell y + \theta_0$, and

$$B(c, \ell) = \frac{1}{2}\sqrt{\alpha(D + c^2 - \sigma\ell^2)}\,, \quad A(c, \ell) = 6\frac{\alpha}{D}B^2\,.$$

They show that this wave is linearly transversely unstable, with respect to long-wave modulation in the transverse direction, when

$$\alpha\sigma\big(4\sigma\ell^2 - 4c^2 - D\big) > 0\,. \tag{2.9}$$

When $\sigma = +1$, the instability region is a parabolic strip in the (c, ℓ)-plane. They also construct an Evans function and confirm the above instability results, and then extend the theory to a class of solitary waves which are biasymptotic to a nontrivial constant at infinity. The Evans function theory is also used to show that there is a finite cutoff of the transverse wavenumber for instability. Numerical simulations, using a multisymplectic Fourier spectral method, confirm the analytical results and then simulate the fully nonlinear solution. Unlike ZK, which has exact cylindrical solitons in 2+1, the 2+1 two-way Boussinesq equation does not have a 3D solution to go to, and indeed blow-up is found in the numerics. In the case of transverse instability, the rigorous theory of Rousset and Tzvetkov [127] can be used to show that transverse instability persists into the nonlinear regime (see Remark 7 in [127]).

The b-KP equation is studied by Chen et al. [46] for transverse instability. The precise equation they use is

$$\Big[(1-\partial_x^2)u_t + (b+1)uu_x + \kappa u_x - bu_x u_{xx} - uu_{xxx}\Big]_x + \sigma u_{yy} = 0\,, \qquad \sigma = \pm 1\,,$$

for scalar-valued $u(x, y, t)$, where b is a real parameter. They present a range of results on transverse instability of both periodic waves and their infinite wavelength limit, depending on the values of b and the speed c of the basic state, and the sign of σ. These results confirm and extend results of Chen and Jin [44].

Bhavna et al. [15, 16] have studied a range of PDEs, that generalise the KP equation, for transverse instability. In [15], they study the $2+1$ rotation-modified KP equation in the form

$$(u_t + 2uu_x - \beta u_{xxx})_x - \gamma u + u_{yy} = 0\,, \quad \gamma > 0,$$

for scalar-value $u(x, y, t)$. It is a $2+1$ generalisation of the Ostrovsky equation, and reduces to the KP equation when $\gamma = 0$. They prove a range of results on transverse stability and instability of periodic travelling waves. See also Bhavna et al. [14]. In Bhavna et al. [16] the Konopelchenko–Dubrovsky (KD) equation is studied in the form

$$u_t = u_{xxx} - 6\rho uu_x + \frac{3}{2}\gamma^2 u^2 u_x - 3v_y + 3\gamma u_x v = 0\,,$$

for $u(x, y, t)$ and $v(x, y, t)$ along with the constraint $u_y = v_x$, where ρ and γ are real parameters. The $1+1$ restriction of this equation is the Gardner equation, and when $\gamma = 0$ it reduces to the KP equation with negative dispersion (KP-II). They give a rigorous analysis of transverse stability and instability of the $1+1$ periodic travelling waves. It would be of interest to extend this theory to a transverse stability analysis of Gardner-equation fronts in the KD equation.

A range of generalisations of KP are discussed in Wang and Li [137]. Most of these have solitary wave solutions of one form or another, but many of these have yet to be studied for transverse instability.

In Pelinovsky and Shimabukuro [120] a class of $2+1$ Dirac equations is studied of the form

$$\begin{aligned} \mathtt{i}(u_t + u_x) + v + u_{yy} &= (\alpha_1|u|^2 + \alpha_2|v|^2)u\,, \\ \mathtt{i}(v_t - v_x) + u + v_{yy} &= (\alpha_2|u|^2 + \alpha_1|v|^2)v\,, \end{aligned} \tag{2.10}$$

for complex-valued functions $u(x, y, t)$ and $v(x, y, t)$, where α_1 and α_2 are real parameters. These equations are closer in spirit to NLS equations since they have two symmetries (translation invariance in space as well as a gauge invariance), but are mentioned here since Dirac-type operators are the main class of PDEs developed in Chap. 3.

The two cases of interest in [120] are $\alpha_2 = 2\alpha_1$ and $\alpha_1 = 0$, $\alpha_2 \neq 0$. The $1+1$ version of (2.10) can be transformed to multisymplectic form (see §3.2 of [35], where it is called the "coupled-mode equation"). The $2+1$ version can be multisymplectified but is not a multisymplectic Dirac operator of the form introduced in Chap. 3.

The Dirac equation with hexagonal potential studied in [120] is

$$\begin{aligned} \mathrm{i}(u_t + v_x) - v_y + u &= (\beta_1|u|^2 + \beta_2|v|^2)u\,, \\ \mathrm{i}(v_t + u_x) + u_y - v &= (\beta_2|u|^2 + \beta_1|v|^2)v\,, \end{aligned} \tag{2.11}$$

with $u(x, y, t)$ and $v(x, y, t)$ complex-valued and β_1, β_2 real parameters. The special case of the Gross–Neveu model is when $\beta_1 = -\beta_2$.

The basic solitary waves studied in [120] are of the form

$$u(x, y, t) = U_\omega(x)\mathrm{e}^{\mathrm{i}\omega t} \quad \text{and} \quad v(x, y, t) = \overline{U_\omega(x)}\mathrm{e}^{\mathrm{i}\omega t}\,.$$

There are parameter values where explicit localised solitary waves exist. In both cases (2.10) and (2.11), they prove analytically that the line solitons are susceptible to instability with respect to periodic transverse perturbations of large periods. The instability is induced by the spatial translation for the massive Thirring model and by the gauge rotation for the Gross–Neveu model. They also observe numerically that the instability holds for the transverse perturbations of any period in the massive Thirring model and exhibits a finite threshold on the period of the transverse perturbations in the Gross–Neveu model.

2.1 Evans Function Applied to Transverse Instability

The Evans function framework will feature prominently in this monograph. The literature on the Evans function is now vast. Hence we will be brief, and give special attention to previous work using the Evans function to analyse transverse linear stability of solitary waves.

The Evans function approach can be illustrated using the ordinary differential equation (ODE) for the spectral problem associated with the linearisation of the KP equation about a solitary wave solution in (2.3). Introduce the vector

$$V(\xi, \lambda, \nu) = \begin{pmatrix} V_1 \\ V_2 \\ V_3 \\ V_4 \end{pmatrix} := \begin{pmatrix} \widehat{v} \\ \widehat{v}_\xi \\ \widehat{v}_{\xi\xi} + a(\xi)\widehat{v} \\ \widehat{v}_{\xi\xi\xi} + (a(\xi)\widehat{v})_\xi \end{pmatrix},$$

and rewrite (2.3) as

$$\frac{d}{d\xi}\begin{pmatrix} V_1 \\ V_2 \\ V_3 \\ V_4 \end{pmatrix} = \begin{bmatrix} 0 & 1 & 0 & 0 \\ -a(\xi) & 0 & 1 & 0 \\ 0 & 0 & 0 & 1 \\ +\sigma\nu^2 & -\lambda & 0 & 0 \end{bmatrix} \begin{pmatrix} V_1 \\ V_2 \\ V_3 \\ V_4 \end{pmatrix},$$

or, abstractly

$$V_\xi = \mathbf{A}(\xi, \lambda, \nu)V, \quad V \in \mathbb{C}^4. \tag{2.12}$$

The Evans function approach is then to study this spectral problem from a dynamical systems viewpoint. As $\xi \to \pm\infty$, the matrix $\mathbf{A}(\xi, \lambda, \nu)$ is a ξ-independent matrix, $\mathbf{A}^\pm(\lambda, \nu)$, since $a(\xi) \to 0$ as $\xi \to \pm\infty$. It is assumed that the eigenvalues of $\mathbf{A}^\pm(\lambda, \nu)$, for a specified subset of (λ, ν) values, are hyperbolic (zero imaginary part). These eigenvalues then determine the decay rate of the solutions of (2.12). The Evans function itself is then a measure of the distance between the stable manifold emerging from $+\infty$ with the unstable manifold emerging from $-\infty$, and when the intersection is non-trivial it signals the existence of an eigenvalue λ, as a function of ν. Further detail on the nature and properties of the Evans function approach will emerge throughout, and in greatest detail in Chap. 6. A history with references of the Evans function theory can be found in Chapters 8–10 of Kapitula and Promislow [86].

A key idea in this monograph is the identification and utility of the symplectic structure of the Evans function systems. This strategy has wide applicability as many Hamiltonian PDEs can be reformulated as multisymplectic.

The first paper to construct an Evans function for the transverse instability problem is Blyuss et al. [19]. It is applied to the spectral problem for the transverse instability of solitary wave solutions (2.8) of the 2+1 two-way Boussinesq equation. The Evans function can be calculated explicitly (using the strategy in Berryman [13] and Alexander et al. [5]), although the calculations can become unwieldy, and Maple was used. The leading-order terms in the Taylor expansion of the Evans function in terms of λ and ν confirm the instability condition (2.9). Further details on this Evans function calculation can be found in Blyuss [18].

Independently, and shortly thereafter, the spectral problem for transverse instability of solitary waves in a $2+1$ cubic NLS model was studied using an Evans function approach by Deconinck et al. [61]. This paper is revisited below in Sect. 2.3

where transverse instability of NLS models are considered, as NLS models require a different strategy.

In Section 4 of Rousset and Tzvetkov [126], an Evans function argument is used to prove the existence of an unstable eigenvalue, providing one of the necessary conditions for their Theorems 1 and 2 on nonlinear instability (see below in Sect. 2.2 for discussion of nonlinear transverse instability).

In Benzoni-Gavage [11] the transverse instability of solitary waves in Korteweg fluids is studied using an Evans function approach. The governing equations in [11] are

$$\begin{aligned} &\partial_t \rho + \operatorname{div}(\rho u) = 0\,, \\ &\partial_t + (u \cdot \nabla) u = \nabla\big(K(\rho)\Delta\rho + \tfrac{1}{2}K'(\rho)|\nabla\rho|^2 - g_0(\rho)\big)\,, \end{aligned} \tag{2.13}$$

where $\rho > 0$ is the fluid density and u is the vector-valued velocity field. The functions $g_0(\rho)$ and $K(\rho)$ are given smooth functions. Arbitrary transverse space dimension in (2.13) is studied, but we will restrict to $2+1$ in the discussion here. It is shown in [11] that this PDE is Hamiltonian as an evolution equation, and this structure is used in the construction of the $1+1$ solitary wave solutions as they are characterised using an energy-momentum variational principle. It is then shown that the variational principle does not give a stability or instability result for the transverse case. The paper [11] then changes tack in §4 and reformulates the spectral problem in an Evans form,

$$\big(B^\sigma(\eta)\Phi\big)_\xi = A^\sigma(\tau,\eta)\Phi\,, \quad \Phi \in \mathbb{C}^6\,,$$

where $B^\sigma(\eta)$ is non-singular, σ is the wavespeed (denoted by c here), η is the transverse Fourier exponent (ν here), and τ is the stability exponent (λ here). After a lengthy analysis of the Evans function, the main result (Theorem 2 in [11]) in that planar (noted as $1+1$ here) solitary waves, that are stable in $1+1$, are spectrally linearly unstable in $2+1$. Paddick [114] follows up on this result and proves, using [128], that the spectral results in [11] are sufficient to conclude nonlinear transverse instability.

Yamazaki [140] studied the transverse instability of $1+1$ solitary waves for the Zakharov–Kuznetsov equation using an Evans function formulation. The paper also proves nonlinear instability, and develops a normal form for bifurcation from the basic solitary wave. However, it is the formulation of the spectral problem in terms of the Evans function that is of interest in this section. The form of the ZK equation studied is

$$u_t + 2uu_x + u_{xxx} + u_{xyy} = 0 \tag{2.14}$$

for scalar-valued $u(x,y,t)$. The $1+1$ solitary wave is

$$Q_c(x) = \frac{3c}{2}\operatorname{sech}^2\left(\frac{1}{2}\sqrt{c}x\right),$$

using notation in [140], where c is the wavespeed. It is first proved that for $c > 0$ this solitary wave is (1) orbitally stable if $0 < L < 2/\sqrt{5c}$, and (2) orbitally unstable if $L > 2/\sqrt{5c}$, where L is the transverse wavelength of the perturbation. It is the critical value $L = 2/\sqrt{5c}$ that generates a bifurcation, and this bifurcation is studied in Pelinovsky [119] and in §7.1 of [140]. The Evans function theory is applied in §3 of [140] to the spectral problem

$$\frac{d}{dx}\left(-\frac{d^2}{dx^2} + c - 2Q_c(x) + a\right)u - \lambda u = 0\,,$$

where $a = n^2/L^2$ with n an integer and L the transverse wavelength. This equation is re-formulated as the first-order system

$$U_x = \mathbf{A}(x, \lambda, a)U\,, \quad U \in \mathbb{C}^3\,,$$

with $U = (u, u_x, u_{xx})$ and

$$\mathbf{A}(x, \lambda, a) = \begin{pmatrix} 0 & 1 & 0 \\ 0 & 0 & 1 \\ -2Q_c{}'(x) - \lambda & c + a - 2Q_c(x) & 0 \end{pmatrix}.$$

With c, λ carefully chosen, the system at infinity is hyperbolic and so this system is in standard form for an Evans function construction. The Evans function, denoted by $D(a, \lambda)$, is found to have a positive eigenvalue for $L > \sqrt{\lambda_c}$, where λ_c is the critical value of a. This spectral result is then used to prove orbital instability for this region of L.

The use of geometric instability criteria, like (1.38), builds on the result of instability of relative equilibria in Hamiltonian ODEs (e.g. [108, 109] and Section 2 of [35]), and the longitudinal instability of solitary waves in 1+1 (e.g. [20, 28, 29, 35, 117]).

2.2 Nonlinear Stability and Instability

Although the emphasis in this monograph is on linear transverse instability, we point out some work on nonlinear stability and instability which builds on the linear theory. A rigorous proof of nonlinear transverse instability for the KdV solitary wave in KP has been proved by Rousset and Tzvetkov [127, 130]. An abstract result, for a class of Hamiltonian evolution equations

$$u_t = J\big(L_0 u + \nabla F(u)\big),$$

where u is scalar-valued, J is skew-symmetric, and L_0 symmetric, is proved in [126]. They prove, in Theorems 1 and 2, with some technical assumptions, that the existence of an unstable transverse eigenmode implies nonlinear orbital instability of the solitary wave. The result is applied to (a) the generalised (polynomial nonlinearity) KP-I equation, (b) a 2 + 1 NLS equation, (c) a 2 + 1 two-way Boussinesq equation (similar to Eq. (2.7) above, with $a = 1$, $b = -1$, and $f(u) = u^2 - u$), (d) the ZK equation, and (e) the KP-BBM equation. The last equation is

$$u_t - u_{txx} + u_x + \left(u^p\right)_x - \partial_x^{-1} u_{yy} = 0\,,$$

with $u(x, y, t)$ scalar-valued and p a natural number. In [128] an abstract instability result for an eigenvalue problem with parameter is proved and applied to transverse linear instability. This simplifies one of the necessary conditions (existence of an unstable eigenvalue) in [126]. As noted above, Paddick [114] follows up on this result and proves nonlinear orbital stability for the 2D Euler–Korteweg model for capillary fluids (2.13).

Taking a different approach, Pelinovsky [119] and Yamazaki [141] develop a normal form theory for the weakly nonlinear case. In [119], the ZK equation is considered, and a first-order ordinary differential equation

$$\frac{db}{dt} = \lambda'(c_*)(c_+ - c_*)b + \gamma |b|^2 b\,, \tag{2.15}$$

where $\lambda'(c_*) > 0$, $\gamma < 0$, and c_* is the critical speed where the line soliton changes stability, c_+ depends on initial data, and b is the complex coordinate of the unstable eigenfunction associated with transverse instability. This normal form can then be used to prove nonlinear stability (when $c_+ < c_*$) and instability (at larger-than-critical speed) of the weakly nonlinear perturbation, in agreement with Yamazaki [140]. Other forerunners of the above weakly nonlinear analysis include Janssen and Rasmussen [82], Rypdal and Rasmussen [131], and Yamazaki [139] for the NLS equation. Yamazaki [141] takes this analysis to the next level by proving the existence of center-stable manifolds around the line solitary waves of the ZK equation. Godey [71], building on the result of Rousset and Tzvetkov [128], proves a simple criterion for transverse linear instability of nonlinear waves. The proof uses spatial dynamics and it is applied to periodic waves of the Davey-Stewartson equation.

2.3 Transverse Instability for NLS-Type Equations

Transverse instability of solitary wave solutions of the cubic NLS equation has been a major area of study, motivated primarily by the appearance of NLS-type equations as models in optics (see §2.1 in Kivshar and Pelinovksy [95] for references). This subject is too large to review here; we will just highlight a few points that resonate with the work in this monograph.

The basic 2 + 1 NLS model with cubic nonlinearity is

$$\mathrm{i}\psi_t + \psi_{xx} + s\psi_{yy} + a|\psi|^2\psi = 0\,, \tag{2.16}$$

with $\psi(x,t)$ complex-valued, and $s = \pm 1$ and $a = \pm 1$. A key theoretical difference between NLS models and KP, ZK, two-way Boussinesq, and related models, is the fact that there is a second prominent symmetry, the gauge invariance with action $\mathrm{e}^{\mathrm{i}\theta}\psi$, for $\theta \in \mathbb{R}$, in addition to the translation invariance in x. Hence the basic $1+1$ solitary wave can be of the form

$$\psi(x,t) = \mathrm{e}^{\mathrm{i}\Omega t} F(x+ct)\,.$$

An example, when $a = +1$ and $c = 0$, is the exact $1 + 1$ bright solitary wave solution

$$\psi = \sqrt{2\Omega} A \mathrm{sech}(\sqrt{B}x)\mathrm{e}^{\mathrm{i}\Omega t}\,.$$

The transverse instability of this class of solutions is the most well studied.

Unlike the variational principal for the KdV solitary wave (critical points of the energy on a single constraint set), the variational principal for NLS solitary waves has two constraints (momentum and charge). Extending this variational principle to $2 + 1$, which would be required for transverse stability analysis, gets vastly more complicated (cf. Bridges [26]).

Zakharov and Rubenchik [143] were first to look at transverse instability of bright solitary waves of cubic NLS with both $s = \pm 1$. They showed that both cases are transversely unstable. In the elliptic case ($s = +1$) the unstable wavenumbers lie in a finite interval $0 < \nu < \nu_c < +\infty$. In the hyperbolic case ($s = -1$), the unstable wavenumbers also lie in a finite interval but the derivation of the cut-off wavenumber is a complicated problem with differing results in the literature.

In follow-up work of Deconinck et al. [61] the hyperbolic case was considered in more detail. The linearisation of (2.16) about the $1 + 1$ bright solitary wave is complex, but can be split into a pair of real equations. A derivation is given in [61], and leads to an eigenvalue problem for λ of the type

$$(L_- - \nu^2)V = \lambda U \quad \text{and} \quad (L_+ - \nu^2)U = -\lambda V\,,$$

where $L_\pm$ are second order ODEs of the form

$$L_- = -\partial_x^2 + 1 - 2\mathrm{sech}^2(x) \quad \text{and} \quad L_+ = -\partial_x^2 + 1 - 6\mathrm{sech}^2(x)\,.$$

In spite of the relative simplicity of this equation, exact solution is not possible, and anomalies occur. Pelinovsky [118] and Deconinck et al. [61] point out that the branch of eigenvalues interacts with the edge of the continuous spectrum. The paper by Deconinck et al. [61] also uses the Evans function to show that edge bifurcations occur in the transverse instability problem.

A proof of nonlinear transverse instability for NLS models is proved in the paper of Rousset and Tzvetkov [126, 128, 130].

A review of results on NLS-type models, up to the beginning of the twenty-first century, is given in Kivshar and Pelinovsky [95]. They include analytical methods and numerical methods, applied to NLS-type models, KdV-type models, and kinks in the Cahn–Hilliard equation. They also review correlation with experimental observations. Results since 2000 are plentiful and too numerous to mention here. A good starting point is the over 500 citations on Google Scholar of [95].

2.4 The Maslov Index of Solitary Waves

The Maslov index has been used in the study of the linear stability of solitary waves for over 30 years. It was pioneered in the work of Jones [84] and Bose and Jones [22]. A numerical framework for computing the Maslov index of solitary waves was introduced in Chardard et al. [41, 42]. Other approaches to the Maslov index of solitary waves were proposed in Chardard [38] and Chen and Hu [43]. The definition used in [40] and [35] was based on a construction of the Maslov index first proposed by Souriau [133]. Recent developments in the use of the Maslov index and its variants, in the stability of solitary waves, include Cox et al. [54], Beck et al. [10], Latushkin and Sukhtaiev [99], Cox et al. [55], and Cornwell et al. [53]. There is now a vast literature on the subject. Recent reviews are given in in Beck [8], Curran [57], Her and Zhong [76], and fundamentals of the Maslov index are given in the books of de Gosson [60] and Long [105].

The Maslov index is sometimes interpreted as a generalisation of the Sturm oscillation theorem, as it can give an indication of how oscillatory a function is (e.g. Pushkar [122], Howard [77], Barutello et al. [7], Cox et al. [55], Beck and Jaquette [9] and references therein). The oscillation perspective is useful to see how the Maslov index works, by counting intersections between Lagrangian planes. For illustration, let V be a symplectic vector space of dimension 4 with symplectic form

$$\Omega(A, B) = \langle \mathbf{J}A, B\rangle \,, \quad A, B \in V \,.$$

Let $\mathbf{P} \in \mathbb{R}^{4\times 2}$ be a fixed Lagrangian plane, and let $\mathbf{Z}(\xi)$ be a path of Lagrangian planes, for some interval of real numbers, say $a < \xi < b$. The Maslov index is the sum of the signed intersections between $\mathbf{P}$ and $\mathbf{Z}(\xi)$ as ξ varies,

$$\mathsf{Maslov}(\mathbf{Z}, \mathbf{P}) = \sum_{a<\xi_0<b} \operatorname{sign} \Gamma(\mathbf{Z}, \mathbf{P}, \xi_0) \,, \tag{2.17}$$

where Γ is a crossing form: for each one-dimensional intersection $\mathbf{Z}(\xi_0) \cap \mathbf{P}$, at $\xi_0 \in \mathbb{R}$, it is

$$\Gamma(\mathbf{Z}, \mathbf{P}, \xi_0) = \Omega\big(\mathbf{Z}'(\xi_0)\beta, \mathbf{Z}(\xi_0)\beta\big), \quad \text{with} \quad \mathbf{Z}'(\xi_0) := \frac{d}{d\xi} Z(\xi)\Big|_{\xi=\xi_0},$$

where $\beta \in \mathbb{R}^2$ determines the linear combination of the columns of $\mathbf{Z}(\xi_0)$ which span the intersection subspace. The formula (2.17) assumes that $\mathbf{Z}(\xi)$ has no intersection with $\mathbf{P}$ at the endpoints and that there exists only a finite number of intersection points. In reality more complex situations arise, and a massive number of variants and generalisations of the Maslov index are now in the literature. See the aforementioned references.

In this monograph, the Maslov index appears in the derivative formula (1.38) indirectly via the transversality index. The transversality index can be expressed in terms of the parity of the Maslov index, a result first proved by Chardard and Bridges [40], and we will record the necessary details needed here in Appendix D.

2.5 Transverse Instability for Water Waves

Many of the model equations discussed above originally arose as approximate equations for water waves. The KP equation is a model for shallow water waves when amplitude, dispersion and transverse variation are in balance, the $2+1$ NLS equation is a model for modulation of Stokes travelling waves in deep water, and the Gardner equation, and its $2+1$ generalisation, arise in the theory of internal waves. Moreover, new model equations for water waves are always emerging. A recent example is the $2+1$ generalised Klein–Gordon equation, first derived by Clamond and Dutykh [49] and studied numerically by Dutykh et al. [66]. It is a model for deep water that has solitary wave solutions, and has been shown to be multisymplectic.

However, this section of the review is about analytical techniques, numerics, and rigorous proofs of transverse instability for the full water wave problem. The most well-known solitary wave in the theory of water waves is the KdV solitary wave in shallow water and its fully nonlinear extension. The fully nonlinear $1+1$ solitary wave is linearly stable to longitudinal perturbations up to almost the highest wave. Longuet-Higgins and Tanaka [106] show that the momentum (equivalently the energy) plotted as a function of wavespeed has a maximum at the change of stability. A schematic of the momentum–wavespeed diagram is shown in Fig. 2.1. Further unstable eigenvalues emerge at higher amplitudes whenever $dI/dc = 0$. A transverse instability emerges also at the maximum point of the $I - c$ diagram; this is shown analytically in Bridges [25]. This result, and the previous results on longitudinal stability when $dI/dc > 0$, show that the classical solitary water wave is remarkably stable to longitudinal perturbations, but susceptible to transverse instability. These properties are borne out by experiments and observations.

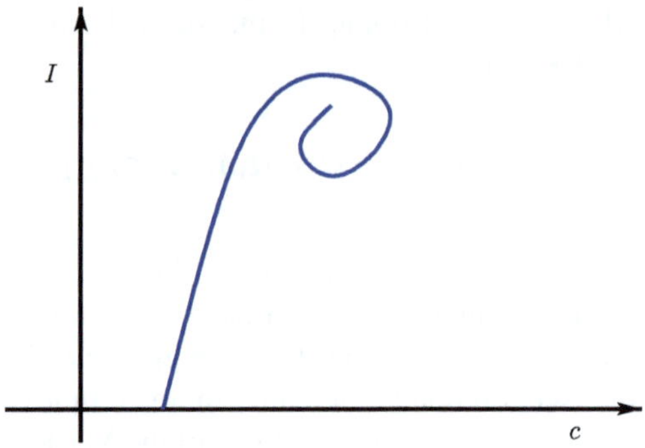

Fig. 2.1 The momentum I plotted as a function of wavespeed c for solitary wave solutions of the water-wave problem

By taking the theory to the next order, and using the method of matched asymptotic expansions, Kataoka and Tsutahara [90] discovered a new transverse instability. They firstly confirm $dI/dc = 0$ to be a transverse instability threshold, but also find parameter regions, at lower amplitude, where the solitary waves are longitudinally stable but transversely unstable. These analytical results are confirmed numerically in Kataoka [88], and a finite wavenumber cut-off is found for transverse instability. These results are extended both analytically and numerically to solitary waves with surface tension by Kim and Akylas [94] and Kim [93]. In Kim and Akylas [94] a theory for transverse instability is developed for the depression solitary waves that bifurcate from the minimum of the phasespeed–wavenumber diagram (denoted by C_{min}) for the linearised gravity–capillary waves. They find numerically that these waves are longitudinally stable but transversely unstable. These results confirm earlier results of Akylas [3] which finds transverse instability using a $2 + 1$ NLS model. His result is extended to transverse instabilities when constant vorticity is added to the flow by Abid et al. [1]. They find that the presence of vorticity influences strongly properties of the basic $1 + 1$ solitary wave, and influences the growth rate of unstable transverse perturbations. This work is further extended to finite depth in Kharif et al. [92]. For pure capillary waves, Murashige and Choi [112] find numerically that pure capillary waves are two-dimensionally stable, but three-dimensionally unstable for almost all wave amplitudes.

Park and Cho [115] performed physical experiments on gravity–capillary solitary waves and observe the transverse instability of free 2D gravity–capillary solitary waves and the resultant formation of three-dimensional gravity–capillary solitary waves. These experimental observations are compared with numerical results based on a model equation that admits gravity–capillary solitary wave solutions near C_{min}. They find that experiments and theory are in good agreement.

The theory for transverse instability of interfacial gravity waves is very similar to the case of pure water waves. Kataoka [87] develops an analytical theory for transverse instability of interfacial solitary waves, and also finds a region of existence of solitary waves where they are transversely unstable but longitudinally stable, much like [89] for pure water waves. Other results on transverse instability of internal gravity waves include Dauxois et al. [59], Varma et al. [136] and references therein. Direct numerical simulation of the dynamics of interfacial gravity–capillary waves in 3D is presented in Li and Wang [103]. They do not address stability directly, but look at some of the 3D structures, such as lump solitary waves, that will emerge from transversely unstable solitary waves.

Abid et al. [1] add vorticity and study the transverse instability of gravity–capillary waves bifurcating from the minimum of the wavespeed–wavenumber diagram, C_{min}, extending results of [3, 94]. They establish that such solitary waves exist in deep water with vorticity, and they find that the presence of vorticity strongly influences the rate of growth of transverse instabilities, and present quantitative numerical results.

The first rigorous result on transverse instability of solitary water waves was given by Groves et al. [74]. They give a proof based on spatial dynamics for transverse instability of gravity–capillary waves in finite depth. They prove that the line solitary wave with large surface tension (Bond number greater than $\frac{1}{3}$) is transversely unstable. Another proof of this transverse instability is given by Pego and Sun [116]. They prove also that the growth rates of the unstable modes are well approximated by the analogous unstable modes in the KP equation with positive dispersion. A major step forward was made by Rousset and Tzvetkov [129] where they prove nonlinear instability for this class of gravity–capillary waves.

The above rigorous results are for gravity–capillary solitary waves. Recently, in Rousset and Sun [125], the pure gravity case is considered. They establish spectral stability as well as the transverse linear asymptotic stability of gravity line solitary waves of the full water wave problem.

Krechetnikov [97] takes the subject of transverse instability of water waves in a different direction, by studying transverse instability of circular solitary waves. He derives a cylindrical KdV equation in shallow water, and shows that the NLS in deep water is replaced by a Gross–Pitaevskii type equation. Properties of both models are studied, when the basic state is axisymmetric, and conditions for transverse instability are derived. Surface tension has a notable effect on the emergence of transverse instability in deep water.

This completes the literature review. It is not comprehensive, but we hope it gives some idea of the vast interest and activity in this area, and gives motivation to the theory presented in this monograph. Transverse stability and instability, for both model equations and the full equations governing water waves, is a flourishing area of research, with many open questions remaining.

Chapter 3
Multisymplectic Wave Equations and Dirac Operators

The theory will be developed for the abstract class of multisymplectic PDEs introduced in (1.26) and repeated here:

$$\mathbf{M}Z_t + \mathbf{K}Z_x + \mathbf{N}Z_y = \nabla S(Z)\,, \quad Z \in \mathbb{R}^N\,, \tag{3.1}$$

where $\mathbf{M}$, $\mathbf{K}$ and $\mathbf{N}$ are $N \times N$ skew-symmetric matrices and ∇S is the gradient of smooth function $S : \mathbb{R}^N \to \mathbb{R}$. In this chapter we show various ways that this class of PDEs can arise, both from transforming Hamiltonian and Lagrangian formulations of well-known PDEs, and from pure geometry.

One of our main results is to connect Dirac operators to multisymplectic Hamiltonian PDEs. The fundamental definition of a Dirac operator, on flat space or flat space-time, is a first order differential operator, whose square is second-order operator, such as a Laplacian or d'Alembertian, with a pre-assigned metric, and the coefficient matrices of the Dirac operator are elements of a Clifford algebra. An example, on a flat space of dimension n, with coordinates, $x_1, \ldots, x_n$, is

$$D = \sum_{j=1}^{n} \mathbf{E}_j \frac{\partial}{\partial x_j}\,,$$

acting on a space of dimension N, where the matrices $\mathbf{E}_j$ are $N \times N$ matrices which satisfy a set of Clifford algebra relations; that is

$$\mathbf{E}_i\mathbf{E}_j + \mathbf{E}_j\mathbf{E}_i = -2\mathcal{R}_{ij}\mathbf{I}_N\,, \quad i, j = 1, \ldots, m\,,$$

where $\mathcal{R}_{ij}$ are entries of the metric. The theory of Clifford algebras relevant here is given in Sect. 3.4 below. A history of Dirac equations is given in the Wikipedia entry [62].

Our contribution here is to develop and apply a class of Dirac operators when the coefficient matrices are each symplectic operators, and also satisfy the Clifford

T. J. Burchell, T. J. Bridges, *Transverse Instability of Solitary Waves*, Mathematical Physics Studies, https://doi.org/10.1007/978-981-95-2046-6_3

algebra relations. Moreover, we identify how the symplectic structures emerge from the total exterior bundle of the manifold M (the base space, which will be taken to be $\mathbb{R}^{1+2}$ herein). This observation is by no means obvious. To show that it is natural, we will approach this construction from two angles. Firstly, we assume we are unfamiliar with Dirac operators, and just attempt to multisymplectify a nonlinear wave equation. This strategy will give some clues that there is a natural intrinsic geometry behind the construction. We will then work backwards from the geometry of space-time.

3.1 Classical Wave Equation as a Dirac Equation

In this section, we start with a classical semi-linear wave equation on $\mathbb{R}^{1,2}$, construct a Lagrangian for which the PDE is the Euler–Lagrange equation and then introduce a Legendre transform to Hamiltonian and multisymplectic form. The starting point is the semi-linear wave equation

$$\Box\phi + V'(\phi) = 0\,, \tag{3.2}$$

where $V(\phi)$ is a given smooth function, $\phi(x, y, t)$ is a scalar-valued function, and $\Box$ is the d'Alembertian operator

$$\Box = \partial_{tt} - \partial_{xx} - \partial_{yy}\,. \tag{3.3}$$

Equation (3.2) is the Euler–Lagrange equation associated with variation of the Lagrangian,

$$\delta \int_D \left[\frac{1}{2}\left(-\phi_t^2 + \phi_x^2 + \phi_y^2\right) + V(\phi)\right] \varpi = 0\,, \tag{3.4}$$

where D is an open set in space-time, $\varpi = \mathrm{d}t \wedge \mathrm{d}x \wedge dy$ is a volume form, and vanishing variations are taken at the boundary of D.

The standard way to generate a first-order system of PDEs from (3.4) is to use a partial (time only) or total (space and time) Legendre transform. We will approach the Legendre transform from the point of view of a *relaxed variational principle* (e.g. Clamond et al. [51] and references therein).

In a relaxed variational principle, new coordinates are introduced to eliminate all derivative terms in the integrand,

$$u = \phi_t\,, \qquad v = \phi_x\,, \qquad w = \phi_y\,,$$

and then introduce these definitions as constraints,

$$\begin{aligned}\mathscr{L} = \int_D \Big[\frac{1}{2}(-u^2 + v^2 + w^2) + V(\phi) - \lambda_1(u - \phi_t) \\ - \lambda_2(v - \phi_x) - \lambda_3(w - \phi_y) \Big] \varpi ,\end{aligned}$$

with $\lambda_1, \lambda_2, \lambda_3$ Lagrange multipliers and new unknowns. This Lagrangian depends on seven variables, and the first variation with respect to all seven variables gives a first-order PDE, although it does not take a tidy form, and is most certainly not in Dirac form. Part of the problem is that there are hidden constraints: $u_x = v_t$, $u_y = w_t$, and $v_y = w_x$. To include these constraints (along with Lagrange multipliers) generates three more unknowns, and it is still not in a tidy form nor in a Dirac form.

At this point we change tactic. The constraints look like the exterior derivative of a one-form. This observation suggests the use of differential forms. For example, $u_x = v_t$ can be interpreted as setting the exterior derivative of the one-form $u\mathrm{d}t + v\mathrm{d}x$ to zero:

$$0 = \mathrm{d}\big(u\mathrm{d}t + v\mathrm{d}x\big) = \big(v_t - u_x\big)\mathrm{d}t \wedge \mathrm{d}x .$$

Fix the basic space-time manifold to be flat, $M = \mathbb{R}^{1,2}$, and take the metric to be

$$g = \mathrm{d}t \otimes \mathrm{d}t - \mathrm{d}x \otimes \mathrm{d}x - \mathrm{d}y \otimes \mathrm{d}y ,$$

with associated matrix

$$\mathcal{R} = \mathrm{diag}(1, -1, -1) . \tag{3.5}$$

This matrix will feature in the Clifford algebra construction below. Introduce the differential forms $u, v, w \in \mathcal{A}(M)$, where $\mathcal{A}^j(M)$ is the space of differential j-forms on $M = \mathbb{R}^{1,2}$,

$$\begin{aligned} u &:= u_1\,\mathrm{d}t + u_2\,\mathrm{d}x + u_3\,\mathrm{d}y, \\ v &:= v_1\,\mathrm{d}x \wedge \mathrm{d}y + v_2\,\mathrm{d}y \wedge \mathrm{d}t + v_3\,\mathrm{d}t \wedge \mathrm{d}x, \\ w &:= w_1 \varpi . \end{aligned} \tag{3.6}$$

The induced inner products for 1, 2 and 3-forms are

$$\begin{aligned} [\![u, u']\!]_1 &= u_1u_1' - u_2u_2' - u_3u_3', \\ [\![v, v']\!]_2 &= v_1v_1' - v_2v_2' - v_3v_3', \\ [\![w, w']\!]_3 &= w_1w_1'. \end{aligned} \tag{3.7}$$

with $u', v', w' \in \mathcal{A}(\mathbb{R}^{1,2})$. Now rewrite the Lagrangian variational principle in (3.4) in terms of differential forms

$$\delta \int_D \left[\frac{1}{2} [\![\mathrm{d}\phi, \mathrm{d}\phi]\!]_1 - V(\phi) \right] \varpi = 0 . \tag{3.8}$$

In the spirit of relaxed variational principles, introduce new variables

$$u = \mathrm{d}\phi \in \mathcal{A}^1(M), \quad v = \mathrm{d}u \in \mathcal{A}^2(M), \quad \text{and} \quad w = \mathrm{d}v \in \mathcal{A}^3(M).$$

The new variational principle on differential form space is

$$\delta \int_D \mathcal{L}(\phi, u, \alpha, v, w) \varpi = 0 ,$$

with Lagrangian density,

$$\mathcal{L}(\phi, u, \alpha, v, w) = \frac{1}{2} [\![u, u]\!]_1 - V(\phi) - [\![\alpha, u - \mathrm{d}\phi]\!]_1 + [\![v, \mathrm{d}\alpha]\!]_2 + [\![w, \mathrm{d}v]\!]_3 ,$$

where $\alpha \in \mathcal{A}^1(M)$, and α, v, w are Lagrange multipliers.

Take the first variation of $\mathcal{L}$ with respect to u

$$\frac{\mathrm{d}}{\mathrm{d}\varepsilon} \left[\int_D \mathcal{L}(\phi, u + \varepsilon \tilde{u}, \alpha, v, w) \varpi \right]_{\varepsilon = 0} = 0,$$

$$\Rightarrow \qquad \int_D [\![\tilde{u}, u - \alpha]\!]_1 \varpi = 0, \tag{3.9}$$

which means that $\alpha = u$, since $\tilde{u}$ is arbitrary. Therefore, $\mathcal{L}$ can be simplified to:

$$\mathcal{L}(\phi, u, v, w) = -\frac{1}{2} [\![u, u]\!]_1 - V(\phi) + [\![u, \mathrm{d}\phi]\!]_1 + [\![v, \mathrm{d}u]\!]_2 + [\![w, \mathrm{d}v]\!]_3 . \tag{3.10}$$

Define $\hat{\phi}$, $\hat{u}$, $\hat{v}$ and $\hat{w}$ as arbitrary variations which are 0 on ∂D. Then by taking first variations of the Lagrangian (3.10) in each coordinate direction we obtain the equations

$$\int_D \left([\![u, \mathrm{d}\hat{\phi}]\!]_1 - \hat{\phi} V'(\phi) \right) \varpi = 0, \tag{3.11}$$

$$\int_D \left([\![\hat{u}, \mathrm{d}\phi]\!]_1 + [\![v, \mathrm{d}\hat{u}]\!]_2 - [\![\hat{u}, u]\!]_1 \right) \varpi = 0, \tag{3.12}$$

$$\int_D \left([\![\hat{v}, \mathrm{d}u]\!]_2 + [\![w, \mathrm{d}\hat{v}]\!]_3 \right) \varpi = 0, \tag{3.13}$$

$$\int_D [\![\hat{w}, \mathrm{d}v]\!]_3 \varpi = 0. \tag{3.14}$$

Introducing the Hodge star operator $\bigstar$ and codifferential δ (e.g. Morita [110]), it is an explicit calculation to verify that, for any k-form α and $(k+1)$-form β,

$$\int_D \left([\![\beta, \mathrm{d}\alpha]\!]_{k+1} - [\![\alpha, \delta\beta]\!]_k \right) \varpi = \int_{\partial D} \alpha \wedge \bigstar\beta\,. \tag{3.15}$$

Hence, if α is 0 on ∂D then

$$\int_D [\![\beta, \mathrm{d}\alpha]\!]_{k+1} \varpi = \int_D [\![\alpha, \delta\beta]\!]_k \varpi. \tag{3.16}$$

Applying this to Eqs. (3.11)–(3.13) gives

$$\int_D \hat{\phi}\left(\delta u - V'(\phi)\right) \varpi = 0, \tag{3.17}$$

$$\int_D [\![\hat{u}, \mathrm{d}\phi + \delta v - u]\!]_1 \varpi = 0, \tag{3.18}$$

$$\int_D [\![\hat{v}, \mathrm{d}u + \delta w]\!]_2 \varpi = 0, \tag{3.19}$$

and so as $\hat{\phi}$, $\hat{u}$, $\hat{v}$ and $\hat{w}$ are arbitrary, Eqs. (3.17)–(3.19) and (3.14) will be satisfied if and only if:

$$\delta u = V'(\phi), \tag{3.20}$$

$$\mathrm{d}\phi + \delta v = u, \tag{3.21}$$

$$\mathrm{d}u + \delta w = 0, \tag{3.22}$$

$$\mathrm{d}v = 0. \tag{3.23}$$

Although not immediately obvious, the left-hand side is in the form of a Dirac operator with both a Clifford algebra structure and, with respect to the appropriate metric, a multisymplectic structure. These properties will emerge after we look more closely at the underlying geometry.

Rewrite this system as a Hamiltonian-type system

$$\mathbf{J}_\partial Z = \mathcal{R}^{(8)} \nabla S(Z), \tag{3.24}$$

where $Z = (\phi, u, v, w)$, $\mathbf{J}_\partial$ is the operator defined as

$$\mathbf{J}_\partial Z = \begin{pmatrix} 0 & \delta & 0 & 0 \\ \mathrm{d} & 0 & \delta & 0 \\ 0 & \mathrm{d} & 0 & \delta \\ 0 & 0 & \mathrm{d} & 0 \end{pmatrix} \begin{pmatrix} \phi \\ u \\ v \\ w \end{pmatrix}, \tag{3.25}$$

S is the scalar-valued function

$$S(Z) = V(\phi) + \frac{1}{2}[\![u, u]\!]_1,$$

and

$$\mathcal{R}^{(8)} = \text{diag}(1, 1, -1, -1, 1, -1, -1, 1)\,.$$

The importance of $\mathcal{R}^{(8)}$ as an induced metric will be made clear in the next section.

We could put $\mathbf{J}_\partial$ into coordinates and investigate the Dirac, Clifford, and multisymplectic structure. However, at this point we shift to a geometric point of view and show that the operator $\mathbf{J}_\partial$ emerges from the geometry of space-time. This point of view will give more structure to $\mathbf{J}_\partial$, and shows how the operator $\mathcal{R}^{(8)}$ arises naturally.

3.2 The Total Exterior Algebra (TEA) Bundle

The TEA bundle of a Riemannian or pseudo-Riemannian manifold is the collection of all differential forms on the manifold, from the zero-forms (functions), one-forms, on up to the n-forms (volume forms), where n is the manifold dimension. Here we consider the case where M is the flat pseudo-Riemannian manifold $\mathbb{R}^{q,p}$, with constant signature metric represented by

$$[\![\mathbf{u}, \mathbf{v}]\!]_1 := \langle \mathcal{R}\mathbf{u}, \mathbf{v} \rangle\,, \tag{3.26}$$

with $\langle \cdot, \cdot \rangle$ a Euclidean inner product, and

$$\mathcal{R} = \text{diag}(\underbrace{1, \cdots, 1}_{\text{q times}}, \underbrace{-1, \cdots, -1}_{\text{p times}})\,. \tag{3.27}$$

The metric $[\![\cdot, \cdot]\!]_1$ induces a metric on each of the spaces $\bigwedge^k(T_x^*M)$. The induced metrics are denoted by

$$[\![\mathbf{u}^{(k)}, \mathbf{v}^{(k)}]\!]_k \text{ for } \mathbf{u}^{(k)}, \mathbf{v}^{(k)} \in \mathcal{A}^k(M)\,, \tag{3.28}$$

where $\mathcal{A}^k(M)$ is the space of mappings from M into $\bigwedge^k(T_x^*M)$. Concatenating these spaces gives the TEA bundle denoted by

$$\mathcal{A}(M) := \bigcup_{k=0}^{n} \mathcal{A}^k(M)\,. \tag{3.29}$$

The Euclidean space representation of the lifted metric on the TEA bundle is

$$[\![\mathbf{u}^{(k)},\mathbf{v}^{(k)}]\!]_k = \left\langle \mathcal{R}^{(r)}\mathbf{u}^{(k)^r},\mathbf{v}^{(k)^r}\right\rangle, \quad \text{with} \quad r = \dim\left(\bigwedge^{k}(T_x^*M)\right), \tag{3.30}$$

where $\mathbf{u}^{(k)^r}$ is the lift to $\mathbb{R}^r$ of $\mathbf{u}^{(k)}$. It is the lifted mapping $\mathcal{R}^{(r)}$ that will show up in the governing equations, generalising (3.24).

The density of the Lagrangian (3.10) can be interpreted as a generalised form of Hamilton's principle and the last three terms,

$$[\![u, \mathrm{d}\phi]\!]_1 + [\![v, \mathrm{d}u]\!]_2 + [\![w, \mathrm{d}v]\!]_3\,, \tag{3.31}$$

can be interpreted as a generalisation of $\mathbf{p}\cdot\mathbf{dq}$. In order to get the ordering right, we introduce the Hodge star operator into this formula. Hence the following differential form is proposed as an abstract generalisation of (3.31),

$$\Theta(Z) = \sum_{k=1}^{n} \mathbf{u}^{(k)} \wedge \bigstar\mathbf{du}^{(k-1)}\,, \tag{3.32}$$

where $\bigstar$ is the Hodge star and d is the exterior derivative.

Not only is this form the natural generalisation of $\mathbf{p}\cdot\mathbf{dq}$ on the cotangent bundle of a Riemannian manifold in classical mechanics; it turns out that the first variation, relative to the induced metric on the TEA bundle, generates a multisymplectic Dirac operator.

In [27] this form was introduced, in the case of a positive definite metric, for the analysis of nonlinear elliptic PDEs, and it was generalised in [34] for the case of general indefinite metric, which generates a hyperbolic partial differential operator. In [33], the TEA bundle approach is embedded in the variational bicomplex. Here we record the results from [34] needed for the case of hyperbolic Dirac operators.

3.3 Variation of Θ and Dirac Operators

The main result of this section is to prove that the first variation of the form (3.32) is

$$\frac{d}{d\varepsilon}\left[\int_D \Theta(Z+\varepsilon\widehat{Z})\varpi\right]_{\varepsilon=0} = \left\langle \begin{pmatrix} 0 & \delta & 0 & \cdots & 0 \\ \mathrm{d} & 0 & \delta & & \vdots \\ 0 & \mathrm{d} & & \ddots & \\ \vdots & & \ddots & & \delta \\ 0 & \cdots & & \mathrm{d} & 0 \end{pmatrix} \begin{pmatrix} u^{(0)} \\ \\ \vdots \\ \\ u^{(n)} \end{pmatrix}, \begin{pmatrix} \hat{u}^{(0)} \\ \\ \vdots \\ \\ \hat{u}^{(n)} \end{pmatrix} \right\rangle,$$

on a simply-connected domain $D \subset M$, where $Z = \left(u^{(0)}, \ldots, u^{(n)}\right)$, $u^{(k)} \in \mathcal{A}^k(M)$. The operator in brackets is precisely $\mathbf{J}_\partial$. Indeed, the canonical form Θ in (3.32) was discovered by working backwards; that is, taking the anti-derivative of $\mathbf{J}_\partial$. We will build on this result to create a family of nonlinear hyperbolic PDEs that are constructed around the Dirac operator $\mathbf{J}_\partial$.

The properties of the Hodge star operator can be used to rewrite the form Θ as the functional

$$\Theta(Z) = \sum_{k=1}^{n} [\![u^{(k)}, \mathrm{d}u^{(k-1)}]\!]_k \varpi ,$$

which is closer in form to the construction (3.31) using the relaxed variational principle. The nonlinear PDEs are constructed by adding a function $S : \mathcal{A}(M) \to \mathbb{R}$ to the Lagrangian density,

$$\int_D \mathcal{L}(Z)\varpi , \quad \mathcal{L}(Z)\varpi = \Theta(Z) - S(Z)\varpi = \sum_{k=1}^{n} [\![u^{(k)}, \mathrm{d}u^{(k-1)}]\!]_k - S(Z)\varpi , \tag{3.33}$$

generalising (3.10).

For variations, take $\widehat{Z} = \left(\hat{u}^{(0)}, \ldots, \hat{u}^{(n)}\right)$ as an arbitrary element in $\mathcal{A}(M)$ which vanishes on ∂D; then

$$0 = \frac{\mathrm{d}}{\mathrm{d}\varepsilon}\left[\int_D \mathcal{L}(Z + \varepsilon\widehat{Z})\varpi\right]_{\varepsilon=0} = \int_D \left[2\vartheta(Z, \widehat{Z}) - \langle\!\langle \widehat{Z}, \mathcal{R}^{(r)}\nabla S(Z)\rangle\!\rangle\right]\varpi , \tag{3.34}$$

where $\mathcal{R}^{(r)}$ is the induced metric on $\mathcal{A}(M)$, and

$$\begin{aligned} \vartheta(Z, W) &= \frac{1}{2}\left[\Theta(Z + W) - \Theta(Z) - \Theta(W)\right] \\ &= \frac{1}{2} \textstyle\sum_{k=1}^{n} \left([\![u^{(k)}, \mathrm{d}v^{(k-1)}]\!]_k + [\![v^{(k)}, \mathrm{d}u^{(k-1)}]\!]_k \right) , \end{aligned}$$

and $W = \left(v^{(0)}, \ldots, v^{(n)}\right)$. ϑ is the bilinear form associated to Θ and $\langle\!\langle \cdot, \cdot \rangle\!\rangle$ denotes the induced pseudo-inner product on the total exterior algebra bundle

$$\langle\!\langle Z, W \rangle\!\rangle = \sum_{k=0}^{n} [\![u^{(k)}, v^{(k)}]\!]_k .$$

This is equivalent to $\langle \mathcal{R}^{(r)} Z, W\rangle$ where $\langle \cdot\,, \cdot\rangle$ denotes the standard Euclidean inner product on $\mathbb{R}^r$, $r = 2^n$. Now, by applying (3.16) we can rewrite (3.34) as:

$$\int_D \left[\sum_{k=1}^{n} \left([\![\hat{u}^{(k-1)}, \delta u^{(k)}]\!]_{k-1} + [\![\hat{u}^{(k)}, \mathrm{d}u^{(k-1)}]\!]_k \right) - \langle\!\langle \widehat{Z}, \mathcal{R}^{(r)} \nabla S(Z) \rangle\!\rangle \right] \varpi = 0$$

$$\Rightarrow \qquad \int_D \langle\!\langle \widehat{Z}, \mathbf{J}_\partial Z - \mathcal{R}^{(r)} \nabla S(Z) \rangle\!\rangle \varpi = 0, \tag{3.35}$$

where $\mathbf{J}_\partial$ here is the generalised version of (3.25) with action

$$\mathbf{J}_\partial Z = \begin{pmatrix} 0 & \delta & 0 & \cdots & 0 \\ \mathrm{d} & 0 & \delta & & \vdots \\ 0 & \mathrm{d} & & \ddots & \\ \vdots & & \ddots & & \delta \\ 0 & \cdots & & \mathrm{d} & 0 \end{pmatrix} \begin{pmatrix} u^{(0)} \\ \\ \vdots \\ \\ u^{(n)} \end{pmatrix}. \tag{3.36}$$

Since $\widehat{Z}$ is arbitrary, we are left with a Hamiltonian-type system of the same form as (3.24),

$$\mathbf{J}_\partial Z = \mathcal{R}^{(r)} \nabla S(Z). \tag{3.37}$$

3.4 Clifford Algebra Structure

The subject of Clifford algebras is quite wide ranging with many dimensions (e.g. Lounesto [107] and Porteous [121] and references therein). Here, we will focus on the Clifford algebra associated with flat space-time and its generalisations, based on $\mathbb{R}^{q,p}$ with diagonal metric

$$\mathcal{R} = \mathrm{diag}(\underbrace{1, \cdots, 1}_{\text{q times}}, \underbrace{-1, \cdots, -1}_{\text{p times}}). \tag{3.38}$$

There are $q + p$ generators, denoted by $\{\mathbf{E}_1, \ldots, \mathbf{E}_{q+p}\}$, and they act on a space of dimension 2^{q+p}. The multiplication operation is

$$\mathbf{E}_i \mathbf{E}_j + \mathbf{E}_j \mathbf{E}_i = -2\mathcal{R}_{ij} \mathbf{I}, \quad i, j = 1, \ldots, q + p.$$

The diagonal entries satisfy $\mathbf{E}_j^2 = -\mathcal{R}_{jj} \mathbf{I}$; that is, the first q satisfy $\mathbf{E}_j^2 = -\mathbf{I}$ and the remaining p satisfy $\mathbf{E}_j^2 = +\mathbf{I}$. This Clifford algebra is denoted by $\mathscr{C}\ell_{q,p}$.

An example is the Clifford algebra $\mathscr{C}\ell_{1,1}$ on the vector space $\mathbb{R}^4$, with metric $\mathcal{R} = \mathrm{diag}(1, -1)$ and generators

$$\mathbf{E}_1 = \begin{pmatrix} 0 & -1 & 0 & 0 \\ 1 & 0 & 0 & 0 \\ 0 & 0 & 0 & -1 \\ 0 & 0 & 1 & 0 \end{pmatrix} \quad \text{and} \quad \mathbf{E}_2 = \begin{pmatrix} 0 & 0 & 1 & 0 \\ 0 & 0 & 0 & -1 \\ 1 & 0 & 0 & 0 \\ 0 & -1 & 0 & 0 \end{pmatrix}.$$

The generators satisfy the multiplication rule

$$\mathbf{E}_i\mathbf{E}_j + \mathbf{E}_j\mathbf{E}_i = -2\mathcal{R}_{ij}\mathbf{I}.$$

This Clifford algebra will be used in the example in Sect. 3.5 below.

The general definition of a Clifford algebra is as follows.

Definition 1 Let V be a vector space over a field K with a quadratic form $Q : V \to K$. In most cases of interest, the field K is either $\mathbb{R}$ or $\mathbb{C}$. A Clifford algebra is a unital associative algebra generated by V, where

$$v^2 = Q(v)1_C \qquad \forall\, v \in V,$$

where 1_C is the multiplicative identity of the algebra. Provided that the characteristic of K is not 2, this expression is equivalent to

$$uv + vu = 2B(u, v)1_C \qquad \forall\, u, v \in V,$$

where $B(u, v) = \frac{1}{2}[Q(u+v) - Q(u) - Q(v)]$ is the bilinear form associated with Q. When V is a finite-dimensional vector space over the reals and Q is non-degenerate, the Clifford algebra is written as $C\ell_{q,p}(\mathbb{R})$ and it is generated by

$$\mathbf{E}_j^2 = -1 \quad \text{for} \quad j = 1, \ldots, q, \qquad \mathbf{E}_j^2 = +1 \quad \text{for} \quad j = p+1, \ldots, p+q\,,$$

where $\{\mathbf{E}_1, \ldots, \mathbf{E}_{p+q}\}$ forms an orthogonal basis for V, i.e. $B\big(\mathbf{E}_i, \mathbf{E}_j\big) = 0$ for $i \neq j$. This latter property can be used to show that

$$\mathbf{E}_i\mathbf{E}_j = -\mathbf{E}_j\mathbf{E}_i\,, \quad \text{for} \quad i \neq j\,.$$

3.5 Example of $\mathbf{J}_\partial$ on $M = \mathbb{R}^{1,1}$

Taking $M = \mathbb{R}^{1,1}$ with $\mathcal{R} = \mathrm{diag}(1, -1)$, volume form $\varpi = \mathrm{d}t \wedge \mathrm{d}x$, $r = 2^2$, and $\square = \partial_{tt} - \partial_{xx}$ as the d'Alembertian results in the system,

$$\mathbf{J}_\partial Z = \mathcal{R}^{(4)}\nabla S(Z)\,, \tag{3.39}$$

where in this case $\mathcal{R}^{(4)} = \mathrm{diag}(1, 1, -1, -1)$. The operator

$$\mathbf{J}_\partial Z = \begin{pmatrix} 0 & \delta & 0 \\ \mathrm{d} & 0 & \delta \\ 0 & \mathrm{d} & 0 \end{pmatrix} \begin{pmatrix} \phi \\ u \\ v \end{pmatrix}$$

can be written in coordinate form (using coordinates $u = u_1 \mathrm{d}t + u_2 \mathrm{d}x,\ v = v_1 \varpi$):

$$\mathbf{J}_\partial \begin{pmatrix} \phi \\ u \\ v \end{pmatrix} = \left[\mathbf{J}_1 \partial_t + \mathbf{J}_2 \partial_x \right] \begin{pmatrix} \phi \\ u_1 \\ u_2 \\ v_1 \end{pmatrix}$$

with

$$\mathbf{J}_1 = \begin{pmatrix} 0 & -1 & 0 & 0 \\ 1 & 0 & 0 & 0 \\ 0 & 0 & 0 & -1 \\ 0 & 0 & 1 & 0 \end{pmatrix}, \qquad \mathbf{J}_2 = \begin{pmatrix} 0 & 0 & 1 & 0 \\ 0 & 0 & 0 & -1 \\ 1 & 0 & 0 & 0 \\ 0 & -1 & 0 & 0 \end{pmatrix}. \tag{3.40}$$

By multiplying these matrices it can be seen that

$$\mathbf{J}_i \mathbf{J}_j + \mathbf{J}_j \mathbf{J}_i = -2\mathcal{R}_{ij} \mathbf{I}_4 .$$

Therefore, they will generate the Clifford algebra $C\ell_{1,1}(\mathbb{R})$ by taking $\mathbf{E}_1 = \mathbf{J}_1$, $\mathbf{E}_2 = \mathbf{J}_2$ and $1_C = \mathbf{I}_4$. The operator $\mathbf{J}_\partial$ can be connected to the d'Alembertian by

$$\mathbf{J}_\partial \circ \mathbf{J}_\partial = (\mathbf{J}_1 \partial_t + \mathbf{J}_2 \partial_x)^2 = (\mathbf{J}_1)^2 \partial_{tt} + (\mathbf{J}_1 \mathbf{J}_2 + \mathbf{J}_2 \mathbf{J}_1) \partial_{tx} + (\mathbf{J}_2)^2 \partial_{xx} = -\Box \otimes \mathbf{I}_4 ,$$

showing that it acts as a Dirac operator.

The operator $\mathbf{J}_1$ is skew-symmetric and $\mathbf{J}_2$ is symmetric. However, multiplication by the operator $\mathcal{R}^{(4)}$ makes them both skew-symmetric. Since $\mathcal{R}^{(4)}$ is an involution, multiplication of (3.39) $\mathcal{R}^{(4)}$ will result in an equation in standard form for a multisymplectic Hamiltonian PDE,

$$\mathbf{M} Z_t + \mathbf{K} Z_x = \nabla S(Z) . \tag{3.41}$$

The gradient of S is now with respect to the standard Euclidean metric on $\mathbb{R}^4$, and the operators

$$\mathbf{M} := \mathcal{R}^{(4)} \mathbf{J}_1 = \begin{pmatrix} 0 & -1 & 0 & 0 \\ 1 & 0 & 0 & 0 \\ 0 & 0 & 0 & 1 \\ 0 & 0 & -1 & 0 \end{pmatrix}, \qquad \mathbf{K} := \mathcal{R}^{(4)} \mathbf{J}_2 = \begin{pmatrix} 0 & 0 & 1 & 0 \\ 0 & 0 & 0 & -1 \\ -1 & 0 & 0 & 0 \\ 0 & 1 & 0 & 0 \end{pmatrix},$$

represent independent symplectic vector spaces on $\mathbb{R}^4$.

3.5.1 Example: Nonlinear Wave Equation on $\mathbb{R}^{1,1}$

An example nonlinear wave equation that can be cast into the multisymplectic form (3.41) is the following coupled second-order semi-linear wave equation,

$$\phi_{tt} - \phi_{xx} + V_\phi = 0 \quad \text{and} \quad v_{tt} - v_{xx} - V_v = 0\,, \tag{3.42}$$

where $V(\phi, v)$ is a given smooth function. These equations arise in applications where solitary wave solutions are of interest, and they will feature in the theory for stability of solitary waves.

Setting

$$u_1 = \phi_t - v_x \quad \text{and} \quad u_2 = \phi_x - v_t\,, \tag{3.43}$$

the four equations (3.42) and (3.43) can be written as a first-order system,

$$\begin{aligned} (u_1)_t - (u_2)_x &= -V_\phi\,, \\ \phi_t - v_x &= u_1\,, \\ \phi_x - v_t &= u_2\,, \\ (u_1)_x - (u_2)_t &= V_v\,, \end{aligned} \tag{3.44}$$

or alternatively as

$$\mathbf{M}\partial_t \begin{pmatrix} \phi \\ u_1 \\ u_2 \\ v \end{pmatrix} + \mathbf{K}\partial_x \begin{pmatrix} \phi \\ u_1 \\ u_2 \\ v \end{pmatrix} = \begin{pmatrix} V_\phi \\ u_1 \\ -u_2 \\ V_v \end{pmatrix}, \tag{3.45}$$

using the same **M**, **K** as in (3.41). Defining the function

$$S(\phi, u_1, u_2, v) := \frac{1}{2}\left[(u_1)^2 - (u_2)^2\right] + V(\phi, v) \tag{3.46}$$

then completes the transformation from the coupled equations (3.42) to a system of the form (3.41). An analysis of the Evans function for an example in this class of nonlinear wave equations in given in [35].

3.6 Construction of $\mathbf{J}_\partial$ on $M = \mathbb{R}^{1,2}$

In this section, we take $M = \mathbb{R}^{1,2}$ and develop the geometry of the TEA bundle and construct the operator $\mathbf{J}_\partial$ and link this with the construction using the relaxed variational principle in Sect. 3.1, which led to the PDE (3.24). The metric has matrix $\mathcal{R} = \mathrm{diag}(1, -1, -1)$ and the 1-, 2- and 3-forms on M are given in (3.6). These can be used to define the action of d and δ in $\mathbf{J}_\partial$ in coordinates leading to

$$\mathbf{J}_\partial = \mathbf{J}_1 \partial_t + \mathbf{J}_2 \partial_x + \mathbf{J}_3 \partial_y \,.$$

The three matrices have entries

$$\mathbf{J}_1 = \begin{pmatrix} 0 & -1 & 0 & 0 & 0 & 0 & 0 & 0 \\ 1 & 0 & 0 & 0 & 0 & 0 & 0 & 0 \\ 0 & 0 & 0 & 0 & 0 & 0 & -1 & 0 \\ 0 & 0 & 0 & 0 & 0 & 1 & 0 & 0 \\ 0 & 0 & 0 & 0 & 0 & 0 & 0 & -1 \\ 0 & 0 & 0 & -1 & 0 & 0 & 0 & 0 \\ 0 & 0 & 1 & 0 & 0 & 0 & 0 & 0 \\ 0 & 0 & 0 & 0 & 1 & 0 & 0 & 0 \end{pmatrix},$$

$$\mathbf{J}_2 = \begin{pmatrix} 0 & 0 & 1 & 0 & 0 & 0 & 0 & 0 \\ 0 & 0 & 0 & 0 & 0 & 0 & -1 & 0 \\ 1 & 0 & 0 & 0 & 0 & 0 & 0 & 0 \\ 0 & 0 & 0 & 0 & 1 & 0 & 0 & 0 \\ 0 & 0 & 0 & 1 & 0 & 0 & 0 & 0 \\ 0 & 0 & 0 & 0 & 0 & 0 & 0 & 1 \\ 0 & -1 & 0 & 0 & 0 & 0 & 0 & 0 \\ 0 & 0 & 0 & 0 & 0 & 1 & 0 & 0 \end{pmatrix} \quad \text{and} \quad \mathbf{J}_3 = \begin{pmatrix} 0 & 0 & 0 & 1 & 0 & 0 & 0 & 0 \\ 0 & 0 & 0 & 0 & 0 & 1 & 0 & 0 \\ 0 & 0 & 0 & 0 & -1 & 0 & 0 & 0 \\ 1 & 0 & 0 & 0 & 0 & 0 & 0 & 0 \\ 0 & 0 & -1 & 0 & 0 & 0 & 0 & 0 \\ 0 & 1 & 0 & 0 & 0 & 0 & 0 & 0 \\ 0 & 0 & 0 & 0 & 0 & 0 & 0 & 1 \\ 0 & 0 & 0 & 0 & 0 & 0 & 1 & 0 \end{pmatrix},$$

which will generate the Clifford algebra $C\ell_{1,2}(\mathbb{R})$ (taking $\mathbf{E}_1 = \mathbf{J}_1$, $\mathbf{E}_2 = \mathbf{J}_2$, $\mathbf{E}_3 = \mathbf{J}_3$ and $1_C = \mathbf{I}_8$) since

$$\mathbf{J}_i \mathbf{J}_j + \mathbf{J}_j \mathbf{J}_i = -2\mathcal{R}_{ij} \mathbf{I}_8 \,.$$

As in the 1+1 case, the operator $\mathbf{J}_\partial$ acts as a Dirac operator

$$\mathbf{J}_\partial \circ \mathbf{J}_\partial = -\Box \otimes \mathbf{I}_8 \,, \qquad \Box = \partial_{tt} - \partial_{xx} - \partial_{yy} \,.$$

Finally, the matrices $\mathcal{R}^{(8)}\mathbf{J}_1$, $\mathcal{R}^{(8)}\mathbf{J}_2$ and $\mathcal{R}^{(8)}\mathbf{J}_3$ are all skew-symmetric, where

$$\mathcal{R}^{(8)} = \mathrm{diag}(1, 1, -1, -1, 1, -1, -1, 1) \,, \tag{3.47}$$

so multiplying both sides of (3.24) by $\mathcal{R}^{(8)}$ will result in an equation in standard form for a multisymplectic Hamiltonian PDE,

$$\mathbf{M}Z_t + \mathbf{K}Z_x + \mathbf{N}Z_y = \nabla S(Z)\,, \quad Z \in \mathbb{X}\,, \tag{3.48}$$

where in this case $\mathbb{X}$ can be taken to be $\mathbb{R}^8$. The structure operators in coordinates are

$$\mathbf{M} := \mathcal{R}^{(8)}\mathbf{J}_1 = \begin{pmatrix} 0 & -1 & 0 & 0 & 0 & 0 & 0 & 0 \\ 1 & 0 & 0 & 0 & 0 & 0 & 0 & 0 \\ 0 & 0 & 0 & 0 & 0 & 0 & 1 & 0 \\ 0 & 0 & 0 & 0 & 0 & -1 & 0 & 0 \\ 0 & 0 & 0 & 0 & 0 & 0 & 0 & -1 \\ 0 & 0 & 0 & 1 & 0 & 0 & 0 & 0 \\ 0 & 0 & -1 & 0 & 0 & 0 & 0 & 0 \\ 0 & 0 & 0 & 0 & 1 & 0 & 0 & 0 \end{pmatrix},$$

$$\mathbf{K} := \mathcal{R}^{(8)}\mathbf{J}_2 = \begin{pmatrix} 0 & 0 & 1 & 0 & 0 & 0 & 0 & 0 \\ 0 & 0 & 0 & 0 & 0 & 0 & -1 & 0 \\ -1 & 0 & 0 & 0 & 0 & 0 & 0 & 0 \\ 0 & 0 & 0 & 0 & -1 & 0 & 0 & 0 \\ 0 & 0 & 0 & 1 & 0 & 0 & 0 & 0 \\ 0 & 0 & 0 & 0 & 0 & 0 & 0 & -1 \\ 0 & 1 & 0 & 0 & 0 & 0 & 0 & 0 \\ 0 & 0 & 0 & 0 & 0 & 1 & 0 & 0 \end{pmatrix},$$

and

$$\mathbf{N} := \mathcal{R}^{(8)}\mathbf{J}_3 = \begin{pmatrix} 0 & 0 & 0 & 1 & 0 & 0 & 0 & 0 \\ 0 & 0 & 0 & 0 & 0 & 1 & 0 & 0 \\ 0 & 0 & 0 & 0 & 1 & 0 & 0 & 0 \\ -1 & 0 & 0 & 0 & 0 & 0 & 0 & 0 \\ 0 & 0 & -1 & 0 & 0 & 0 & 0 & 0 \\ 0 & -1 & 0 & 0 & 0 & 0 & 0 & 0 \\ 0 & 0 & 0 & 0 & 0 & 0 & 0 & -1 \\ 0 & 0 & 0 & 0 & 0 & 0 & 1 & 0 \end{pmatrix}$$

form the basis for three independent symplectic structures on $\mathbb{R}^8$.

3.6.1 Hermitian Representation in the $2+1$ Case

Another perspective on the multisymplectic formulation is to rewrite (3.48) in complex coordinates. Let

$$Z = (Z_1, Z_2, \ldots, Z_8),$$

and write out Eq. (3.48) in real coordinates. To complexify, define

$$A_1 = Z_1 + \mathrm{i}Z_5, \quad A_2 = Z_2 + \mathrm{i}Z_8, \quad A_3 = Z_3 + \mathrm{i}Z_4, \quad \text{and } A_4 = Z_6 + \mathrm{i}Z_7.$$

Then differentiating, the equation $\mathbf{J}_\partial Z = 0$ becomes

$$\begin{aligned}
\frac{\partial A_1}{\partial t} + \mathrm{i}\frac{\partial A_4}{\partial x} + \frac{\partial A_4}{\partial y} &= 0, \\
\frac{\partial A_2}{\partial t} - \frac{\partial A_3}{\partial x} + \mathrm{i}\frac{\partial A_3}{\partial y} &= 0, \\
\frac{\partial A_3}{\partial t} - \frac{\partial A_2}{\partial x} - \mathrm{i}\frac{\partial A_2}{\partial y} &= 0, \\
\frac{\partial A_4}{\partial t} - \mathrm{i}\frac{\partial A_1}{\partial x} + \frac{\partial A_1}{\partial y} &= 0,
\end{aligned}$$

or

$$\begin{pmatrix} A_1 \\ A_2 \\ A_3 \\ A_4 \end{pmatrix}_t + \begin{bmatrix} 0 & 0 & 0 & \mathrm{i} \\ 0 & 0 & -1 & 0 \\ 0 & -1 & 0 & 0 \\ -\mathrm{i} & 0 & 0 & 0 \end{bmatrix} \begin{pmatrix} A_1 \\ A_2 \\ A_3 \\ A_4 \end{pmatrix}_x + \begin{bmatrix} 0 & 0 & 0 & 1 \\ 0 & 0 & \mathrm{i} & 0 \\ 0 & -\mathrm{i} & 0 & 0 \\ 1 & 0 & 0 & 0 \end{bmatrix} \begin{pmatrix} A_1 \\ A_2 \\ A_3 \\ A_4 \end{pmatrix}_y = \begin{pmatrix} 0 \\ 0 \\ 0 \\ 0 \end{pmatrix}.$$

This gives a Hermitian representation of the multisymplectic operator $\mathbf{J}_\partial$. It can be made to be variational by multiplying the equation by i.

3.7 Example: Vector-Valued Wave Equation in $\mathbb{R}^{2,1}$

In this section we introduce a set of nonlinear wave equations which can be cast into the form (3.48) in $2+1$. These equations were constructed as an example of a vector-valued nonlinear wave equation with an explicit solitary wave solution, and is used as an example in the instability theory.

Let $\phi(x, y, t)$ be scalar-valued and let $\mathbf{v}(x, y, t)$ be an $\mathbb{R}^3$-valued vector that satisfies the set of four coupled nonlinear wave equations

$$\begin{aligned} \phi_{tt} - \phi_{xx} - \phi_{yy} + V_\phi &= 0\,, \\ (v_1)_{tt} - (v_1)_{xx} - (v_1)_{yy} + V_{v_1} &= 0\,, \\ (v_2)_{tt} - (v_2)_{xx} - (v_2)_{yy} - V_{v_2} &= 0\,, \\ (v_3)_{tt} - (v_3)_{xx} - (v_3)_{yy} - V_{v_3} &= 0\,, \end{aligned} \tag{3.49}$$

where $V(\phi, v_1, v_2, v_3)$ is a given smooth function. The Dirac formulation or multisymplectic formulation can be derived using the relaxed variational principle. However, it follows along similar lines to the scalar case in Sect. 3.1 and so we just sketch the construction.

Introduce auxiliary variables

$$\begin{aligned} u_1 &= \phi_t + (v_2)_y - (v_3)_x\,, & u_2 &= \phi_x - (v_1)_y - (v_3)_t\,, \\ u_3 &= \phi_y + (v_1)_x + (v_2)_t\,, & w &= (v_1)_t + (v_2)_x + (v_3)_y\,. \end{aligned}$$

Combining these four equations with (3.49) results in

$$\begin{aligned} (u_1)_t - (u_2)_x - (u_3)_y &= -V_\phi\,, \\ \phi_t + (v_2)_y - (v_3)_x &= u_1\,, \\ \phi_x - (v_1)_y - (v_3)_t &= u_2\,, \\ \phi_y + (v_1)_x + (v_2)_t &= u_3\,, \\ (u_2)_y - (u_3)_x + w_t &= -V_{v_1}\,, \\ (u_1)_y - (u_3)_t + w_x &= -V_{v_2}\,, \\ (u_1)_x - (u_2)_t - w_y &= V_{v_3}\,, \\ (v_1)_t + (v_2)_x + (v_3)_y &= w\,. \end{aligned} \tag{3.50}$$

Using the same **M**, **K**, **N** as in (3.48), this can be rewritten as

$$\mathbf{M}\partial_t \begin{pmatrix} \phi \\ u_1 \\ u_2 \\ u_3 \\ v_1 \\ v_2 \\ v_3 \\ w \end{pmatrix} + \mathbf{K}\partial_x \begin{pmatrix} \phi \\ u_1 \\ u_2 \\ u_3 \\ v_1 \\ v_2 \\ v_3 \\ w \end{pmatrix} + \mathbf{N}\partial_y \begin{pmatrix} \phi \\ u_1 \\ u_2 \\ u_3 \\ v_1 \\ v_2 \\ v_3 \\ w \end{pmatrix} = \begin{pmatrix} V_\phi \\ u_1 \\ -u_2 \\ -u_3 \\ V_{v_1} \\ V_{v_2} \\ V_{v_3} \\ w \end{pmatrix}, \tag{3.51}$$

and so with the function S defined as

$$S(\phi, u_1, u_2, u_3, v_1, v_2, v_3, w) = \frac{1}{2}\left[(u_1)^2 - (u_2)^2 - (u_3)^2 + w^2\right] + V(\phi, v_1, v_2, v_3)\,, \tag{3.52}$$

this system is in the form of (3.48). In Chap. 9 it is shown that this vector-valued nonlinear wave equation has explicit solitary wave solutions, and it is used to illustrate the instability theory.

3.8 Example: Multisymplectifying the KP Equation

The classical KP equation, generalised to have an arbitrary nonlinearity is

$$(u_t + f(u)_x + u_{xxx})_x + \sigma u_{yy} = 0\,, \tag{3.53}$$

for the scalar-valued function $u(x, y, t)$, where $\sigma = \pm 1$, and $f(u)$ can be any smooth function.

We consider the KP equation here, as it is not natural for multisymplectic structure since **M**, **K** and **N** have different ranks. This property is due to the fact that it is first order in t-derivatives, second order in y-derivatives, and fourth order in x-derivatives. Nevertheless, it can be multisymplectified, and this structure has turned out to be useful in a number of contexts: construction of periodic patterns [23], construction of a variational principle for solitary waves of KP [25], development of multisymplectic numerical schemes for simulating KP [37, 45, 83, 102, 104, 138], and phase dynamics of multiphase wavetrains [123, 124]. Moreover, as noted in Chap. 2, the KP equation is also one of the fundamental equations with a transverse instability.

Given a PDE with a Lagrangian, the relaxed variational principle of [51] and Sect. 3.1 can always be used effectively to derive a multisymplectic structure. An alternative approach is to construct the multisymplectic form formula on a jet bundle space. This latter approach for multisymplectifying KP is developed in Liu and Qin [104], and is used to design conservative numerical schemes in [104], as well as in Jiang et al. [83] and Li and Sun [102]. In coordinates the structure in [104] looks different from that derived here as the phase space has 10 dimensions. The structure developed below and in Appendix A is equivalent, but is improved by reducing the phase space dimension to 8, with the x-direction structure operator (**K** below) non-degenerate.

First, we establish that the KP equation has a natural Lagrangian formulation, once a potential for u is introduced,

$$u = \phi_{xx}\,.$$

In terms of ϕ, the KP equation is

$$\phi_{txxx} + f\,(\phi_{xx})_{xx} + \phi_{xxxxxx} + \sigma\phi_{xxyy} = 0\,. \tag{3.54}$$

A proposed Lagrangian density is

$$L = \frac{1}{2}\phi_{tx}\phi_{xx} + F\,(\phi_{xx}) - \frac{1}{2}\phi_{xxx}^2 + \frac{1}{2}\sigma\phi_{xy}^2\,, \tag{3.55}$$

where $F(u) = \int_0^u f(s)\,\mathrm{d}s$ is an antiderivative. The variational principle

$$\delta\int_{t_1}^{t_2}\int_{x_1}^{x_2}\int_{y_1}^{y_2} L\,\mathrm{d}x\mathrm{d}y\mathrm{d}t = 0\,,$$

on the cube $\mathcal{C} := [x_1, x_2]\times[y_1, y_2]\times[t_1, t_2]\subset\mathbb{R}^3$, with fixed endpoint conditions on the ϕ-variations, recovers the governing equation (3.54).

Here we will just record the multisymplectic structure that follows from (3.55), and in Appendix A a derivation using the relaxed variational principle is given. Introduce new coordinates

$$\phi = q_1\,,\quad v = \phi_x = (q_1)_x\,,\quad u = \phi_{xx} = p_3\,,\quad w = \phi_{xxx} = q_3 = (p_3)_x\,, \tag{3.56}$$

along with the constraints

$$(p_1)_x = 0 \quad\text{and}\quad (p_3)_y - (p_4)_x = 0\,, \tag{3.57}$$

and the definitions

$$\begin{aligned} p_1 &= -\frac{1}{2}(p_3)_t - (p_2)_x - \sigma(p_4)_y\,,\\ p_2 &= \frac{1}{2}(q_2)_t + (q_3)_x - (q_4)_y + f(p_3)\,,\\ p_4 &= (q_2)_y + \frac{1}{\sigma}(q_4)_x\,. \end{aligned} \tag{3.58}$$

These coordinates are chosen because they result in a non-degenerate canonical symplectic structure in the x-direction. The above eight equations form a multisymplectic Hamiltonian PDE in the coordinates $(\mathbf{q}, \mathbf{p})\in\mathbb{R}^8$, of the form

$$\mathbf{M}Z_t + \mathbf{K}Z_x + \mathbf{N}Z_y = \nabla S(Z)\,,\quad Z\in\mathbb{R}^8\,,$$

with

$$S(Z) = p_1q_2 + p_2p_3 + \frac{1}{2}\sigma(p_4)^2 - \frac{1}{2}(q_3)^2 - F(p_3)\,,$$

and structure matrices,

$$\mathbf{M} = \begin{bmatrix} 0&0&0&0&0&0&0&0 \\ 0&0&0&0&0&0&-\frac{1}{2}&0 \\ 0&0&0&0&0&0&0&0 \\ 0&0&0&0&0&0&0&0 \\ 0&0&0&0&0&0&0&0 \\ 0&0&0&0&0&0&0&0 \\ 0&\frac{1}{2}&0&0&0&0&0&0 \\ 0&0&0&0&0&0&0&0 \end{bmatrix}, \quad \mathbf{N} = \begin{bmatrix} 0&0&0&0&0&0&0&0 \\ 0&0&0&0&0&0&0&-\sigma \\ 0&0&0&0&0&0&0&0 \\ 0&0&0&0&0&0&1&0 \\ 0&0&0&0&0&0&0&0 \\ 0&0&0&0&0&0&0&0 \\ 0&0&0&-1&0&0&0&0 \\ 0&\sigma&0&0&0&0&0&0 \end{bmatrix},$$

and

$$\mathbf{K} = \begin{bmatrix} 0&0&0&0&-1&0&0&0 \\ 0&0&0&0&0&-1&0&0 \\ 0&0&0&0&0&0&-1&0 \\ 0&0&0&0&0&0&0&-1 \\ 1&0&0&0&0&0&0&0 \\ 0&1&0&0&0&0&0&0 \\ 0&0&1&0&0&0&0&0 \\ 0&0&0&1&0&0&0&0 \end{bmatrix}.$$

The matrices **M** and **N** are singular, preventing a Dirac structure. However, with the above choice of coordinates the structure matrix **K** is both non-degenerate and a canonical symplectic operator on $\mathbb{R}^8$. Therefore steady and y-independent solutions are orbits of the canonical Hamiltonian system on $\mathbb{R}^8$

$$\mathbf{K}Z_x = \nabla S(Z), \quad Z = \begin{pmatrix} \mathbf{q} \\ \mathbf{p} \end{pmatrix} \in \mathbb{R}^8 .$$

The derivation of the above multisymplectic formulation of KP, starting with (3.55) and using the relaxed variational principle, is given in Appendix A.

3.9 Example: Generalised Klein–Gordon for Water Waves

Another equation that models water waves, recently derived and shown to be multi-symplectic by Clamond and Dutykh [49] and Dutykh et al. [66], is the generalised Klein–Gordon equation. It is derived by assuming an exponential dependence of the velocity field in the vertical direction of the form $\exp(\kappa(z - \eta))$, where κ is a free parameter representing an inverse length scale, z is the vertical coordinate, and $\eta(x, y, t)$ is the free surface elevation. Factoring out the exponential leaves a velocity potential, $\phi(x, y, t)$, that depends only on the horizontal coordinates (x, y)

and time. Using a relaxed variational principle, the following approximate equations are derived in [49, 66]:

$$\begin{aligned} \eta_t + \tfrac{1}{2}\kappa^{-1}\Delta\phi - \tfrac{1}{2}\kappa\phi &= \tfrac{1}{2}\phi\left[\Delta\eta + \kappa(\nabla\eta)^2\right] \\ \phi_t + g\eta &= -\tfrac{1}{2}\nabla\cdot\left[\phi\nabla\phi - \kappa\phi^2\nabla\eta\right]. \end{aligned} \tag{3.59}$$

The parameter g (gravitational constant) is positive, and κ is a free positive parameter. The operators are

$$\Delta\phi := \frac{\partial^2\phi}{\partial x^2} + \frac{\partial^2\phi}{\partial y^2}, \quad \nabla\phi := (\phi_x, \phi_y), \quad \text{and} \quad \nabla\cdot =: \text{divergence}.$$

This pair of equations is called a generalised Klein–Gordon equation in [66] because linearisation about the trivial solution, and elimination of ϕ, leads to a Klein–Gordon equation

$$\eta_{tt} - \frac{g}{2\kappa}\Delta\eta + \frac{1}{2}g\kappa\eta = 0.$$

The pair of Eqs. (3.59) can be written as a first-order system of PDEs as follows:

$$\begin{aligned} 2\kappa\eta_t + \nabla\cdot\mathbf{u} &= \kappa^2\phi - \kappa\mathbf{u}\cdot\boldsymbol{\alpha}, \\ -2\kappa\phi_t - \nabla\cdot\boldsymbol{\gamma} &= 2\kappa g\eta, \\ -\nabla\phi &= -\mathbf{u} - \kappa\phi\boldsymbol{\alpha}, \\ \nabla\eta &= \boldsymbol{\alpha}, \\ \mathbf{0} &= \boldsymbol{\gamma} - \kappa\phi\mathbf{u}, \end{aligned} \tag{3.60}$$

where $\boldsymbol{\alpha} = (\alpha_1, \alpha_2)$ and $\boldsymbol{\gamma} = (\gamma_1, \gamma_2)$. This system is multisymplectic and of the form (3.1) with coordinates

$$Z = (\phi, \eta, u_1, u_2, \gamma_1, \gamma_2, \alpha_1, \alpha_2)^T \in \mathbb{R}^8,$$

Hamiltonian function

$$S(Z) = \boldsymbol{\alpha}\cdot\boldsymbol{\gamma} + \kappa g\eta^2 + \frac{1}{2}(\kappa\phi)^2 - \kappa\phi\mathbf{u}\cdot\boldsymbol{\alpha} - \frac{1}{2}\mathbf{u}\cdot\mathbf{u},$$

and structure matrices

$$\mathbf{M} = 2\kappa\big(\mathbf{e}_1 \otimes \mathbf{e}_2 - \mathbf{e}_2 \otimes \mathbf{e}_1\big)\,,$$
$$\mathbf{K} = \mathbf{e}_1 \otimes \mathbf{e}_3 - \mathbf{e}_3 \otimes \mathbf{e}_1 + \mathbf{e}_5 \otimes \mathbf{e}_2 - \mathbf{e}_2 \otimes \mathbf{e}_5\,,$$
$$\mathbf{N} = \mathbf{e}_1 \otimes \mathbf{e}_4 - \mathbf{e}_4 \otimes \mathbf{e}_1 + \mathbf{e}_6 \otimes \mathbf{e}_2 - \mathbf{e}_2 \otimes \mathbf{e}_6\,,$$

where $\mathbf{e}_j$, $j = 1, \ldots, 8$ are the standard unit vectors in $\mathbb{R}^8$.

The system (3.59) has $1 + 1$ periodic and solitary waves. The latter, when embedded in $2+1$ would be candidates for the theory in this monograph. At present these solutions are only known numerically, and so the key determinant (1.40) would have to be computed numerically.

Clamond and Dutykh and co-authors have discovered a range of other equations that model water waves, and they have shown that these models have a multisymplectic structure. Examples are the multisymplectic structure of Boussinesq-type systems in Durn et al. [64, 65], multisymplectic structure of the Serre–Green–Naghdi equation that is a model for shallow water hydrodynamics in Chhay et al. [47], and an extension of the Serre–Green–Naghdi equation to two-layer internal waves by Clamond and Dutykh [50]. These multisymplectic formulations are in $1 + 1$, but in principle could be extended to $2 + 1$.

Chapter 4
Solitary Wave Solutions and Their Properties

In this chapter we look at the basic class of solitary waves whose stability will be investigated. The existence of solitary waves will be assumed, but we will need to know their geometric properties and their dependence on parameters.

The starting point for the theory is the class of nonlinear wave equations,

$$\mathbf{M}Z_t + \mathbf{K}Z_x + \mathbf{N}Z_y = \nabla S(Z), \quad Z \in \mathbb{X}, \tag{4.1}$$

where $\mathbb{X}$ is the phase space, and will be taken to be $\mathbb{R}^8$. It is the lowest phase space dimension that is interesting in $2+1$. The most important example will be the case where $\mathbf{M}Z_t + \mathbf{K}Z_x + \mathbf{N}Z_y$ is a multi-symplectic Dirac operator (cf. Sect. 3.6). The phase space in that case is interpreted as the total exterior algebra bundle on $\mathbb{R}^{1,2}$, which can be identified with $\mathbb{R}^8$ with standard inner product denoted by $\langle \cdot, \cdot \rangle$. The metric on $\bigwedge(\mathbb{R}^{1,2})$ is different but it can be absorbed into the operators **M**, **K**, and **N** as noted in Sect. 3.6, making $\mathbb{R}^8$ the operational phase space.

In this chapter, where the system (4.1) may have degenerate structure operators, the only requirements we make on the left-hand side of (4.1) is that the three matrices **M**, **K**, and **N** be skew-symmetric and constant matrices, and the linear combination

$$\mathbf{J}(c, \ell) := \mathbf{K} + c\mathbf{M} + \ell\mathbf{N} \tag{4.2}$$

be invertible and skew-symmetric for an open set of points $(c, \ell) \in \mathscr{C} \subset \mathbb{R}^2$. With this property, we can define a symplectic vector space $(\mathbb{X}, \mathbf{\Omega})$ with symplectic form

$$\mathbf{\Omega}(U, V) := \langle \mathbf{J}(c, \ell)U, V \rangle, \quad \forall U, V \in \mathbb{X}, \quad (c, \ell) \in \mathscr{C}\,. \tag{4.3}$$

Throughout $\langle \cdot, \cdot \rangle$ will be the inner product on $\mathbb{X}$. The class of Eqs. (4.1) and the symplecticity of $(\mathbb{X}, \mathbf{\Omega})$ form hypothesis (**H1**).

T. J. Burchell, T. J. Bridges, *Transverse Instability of Solitary Waves*, Mathematical Physics Studies, https://doi.org/10.1007/978-981-95-2046-6_4

Since Eq. (4.1) is variational, with Lagrangian density

$$L = \frac{1}{2}\langle Z, \mathbf{M}Z_t\rangle + \frac{1}{2}\langle Z, \mathbf{K}Z_x\rangle + \frac{1}{2}\langle Z, \mathbf{N}Z_y\rangle - S(Z)\,, \tag{4.4}$$

it is natural to appeal to variational methods for the existence of solitary waves. The most well known variational principle in $1 + 1$ is to characterise solitary waves as critical points of the energy (Hamiltonian) on level sets of the momentum. In $1 + 1$ the classical Hamiltonian function and momentum, in terms of the multisymplectic variables, are

$$H(Z) = \int_{-\infty}^{+\infty} \left(S(Z) - \frac{1}{2}\langle Z, \mathbf{K}Z_x\rangle \right) \mathrm{d}x \quad \text{and} \quad I(Z) = \int_{-\infty}^{+\infty} \frac{1}{2}\langle Z, \mathbf{M}Z_x\rangle \, \mathrm{d}x\,,$$

with the wavespeed c as a Lagrange multiplier and Lagrangian necessary condition

$$\delta \int_{-\infty}^{+\infty} \left(S(Z) - \frac{1}{2}\langle Z, \mathbf{K}Z_x\rangle - \frac{1}{2}c\,\langle Z, \mathbf{M}Z_x\rangle \right) \mathrm{d}x = 0\,,$$

with fixed endpoints on the variations of Z. One of the interesting features of this variational principle is that $d\mathcal{I}/dc$ carries information about critical point type, where $\mathcal{I}$ is the momentum evaluated on a solitary wave. This is sometimes called the Vakhitov–Kolokolov approach (cf. Berkolaiko et al. [12]), and is a special case of the Grillakis–Shatah–Strauss theory [72, 73].

Although $2 + 1$ generalisations of this variational principle are not known, the multisymplectic structure gives a natural $2+1$ variational principle by including the y-direction symplectic structure on the same footing as the t-direction symplectic structure. This strategy leads to a variational principle with two parameters, the wavespeed c and transverse parameter ℓ, which measures the skewness of the direction of the solitary wave, that together will give information about spectral stability of the solitary wave.

4.1 The Form of the Solitary Wave Solution

The basic class of solitary waves is taken to be a line solitary wave

$$Z(x, y, t) := \widehat{Z}(\xi) \quad \text{with} \quad \xi = x + ct + \ell y + \xi_0\,, \tag{4.5}$$

where $c, \ell \in \mathscr{C}$ and ξ_0 is an arbitrary phase shift. This solution is a generalised relative equilibrium on the group of translations $x \mapsto x + s$ for all $s \in \mathbb{R}$. It is generalised in the sense that the flow on the group is a linear function of both t and y,

$$x \mapsto x + ct + \ell y\,.$$

See §2–§3 of [35] for the characterisation of solitary waves as relative equilibria in the multisymplectic setting. It is assumed that a solitary wave solution exists with the following properties (see Hypothesis **H2** below for a summary).

It is assumed that the solitary wave solution has the following asymptotic properties,

$$\lim_{\xi\to\pm\infty} \|\widehat{Z}(\xi)\| = 0\,, \tag{4.6}$$

with the convergence exponential.

The line solitary wave is a solution of the ODE

$$\mathbf{J}(c,\ell)\widehat{Z}_\xi = \nabla S(\widehat{Z}). \tag{4.7}$$

Expanding $\mathbf{J}(c,\ell)$, this is

$$\nabla S(\widehat{Z}) = \mathbf{K}\widehat{Z}_\xi + c\mathbf{M}\widehat{Z}_\xi + \ell\mathbf{N}\widehat{Z}_\xi\,. \tag{4.8}$$

The product of a skew-symmetric matrix with $\frac{d}{d\xi}$ is a symmetric operator and so formally the gradient of a quadratic functional. Hence the equation governing solitary waves (4.8) can be characterised as a constrained critical point problem. Define the functionals

$$\begin{aligned} H(\widehat{Z}) &= \int_{-\infty}^{+\infty}\left[S(\widehat{Z}) - \frac{1}{2}\langle\mathbf{K}\widehat{Z}_\xi, \widehat{Z}\rangle\right] \mathrm{d}\xi\,,\\ I_1(\widehat{Z}) &= \int_{-\infty}^{+\infty}\frac{1}{2}\langle\mathbf{M}\widehat{Z}_\xi, \widehat{Z}\rangle\, \mathrm{d}\xi\,,\\ I_2(\widehat{Z}) &= \int_{-\infty}^{+\infty}\frac{1}{2}\langle\mathbf{N}\widehat{Z}_\xi, \widehat{Z}\rangle\, \mathrm{d}\xi\,. \end{aligned} \tag{4.9}$$

Then the ODE (4.7) or (4.8) is the Lagrange necessary condition for critical points of $H(\widehat{Z})$ restricted to level sets of the functionals $I_1(\widehat{Z})$ and $I_2(\widehat{Z})$, with Euler–Lagrange equation

$$0 = \nabla H(\widehat{Z}) - c\nabla I_1(\widehat{Z}) - \ell\nabla I_2(\widehat{Z}) = \nabla S(\widehat{Z}) - \mathbf{K}\widehat{Z}_\xi - c\mathbf{M}\widehat{Z}_\xi - \ell\mathbf{N}\widehat{Z}_\xi\,. \tag{4.10}$$

This construction leads to a two-parameter family of solitary waves parameterised by Lagrange multipliers c, ℓ, denoted by $\widehat{Z}(\xi, c, \ell)$. Evaluating the functionals I_1 and I_2 on the two-parameter family of solitary waves induces functions $I_1(c,\ell)$ and $I_2(c,\ell)$ (using the same symbols I_1 and I_2 for both the functional and its evaluation).

This two-parameter family is called non-degenerate when

$$\det \begin{bmatrix} \frac{\partial I_1}{\partial c} & \frac{\partial I_1}{\partial \ell} \\ \frac{\partial I_2}{\partial c} & \frac{\partial I_2}{\partial \ell} \end{bmatrix} \neq 0\,. \tag{4.11}$$

This determinant and its entries will feature prominently in the spectral stability analysis.

Hypothesis (**H2**) summarises the above. It assumes that there exists a solitary wave solution of the form (4.5) and (4.6) satisfying (4.7) and (4.11), and it is a sufficiently smooth function of ξ, c and ℓ for all $\xi \in \mathbb{R}$ and $(c, \ell) \in \mathscr{C}$.

4.2 Reversibility in *x* and *y*

When the PDE has a reversibility symmetry, this symmetry can be inherited by the solitary wave solution and hence simplify the stability theory. In this section we define what it means to be reversible in the x- and y-directions. The governing equations (3.48) can also have a t-direction reversibility, that is defined similarly, but will not be needed in the theory herein.

Equation (3.48) is said to be $x-$reversible if there exists a reversor $\mathbf{R}_x$ acting on $\mathbb{X}$ satisfying

$$\begin{aligned} &\mathbf{R}_x\mathbf{M} = \mathbf{M}\mathbf{R}_x\,, \quad \mathbf{R}_x\mathbf{K} = -\mathbf{K}\mathbf{R}_x\,, \quad \mathbf{R}_x\mathbf{N} = \mathbf{N}\mathbf{R}_x\,, \quad \text{and} \\ &S(\mathbf{R}_x Z) = S(Z), \quad \forall\, Z \in \mathbb{X}. \end{aligned} \tag{4.12}$$

A "reversor" is a matrix that is both an involution and isometry

$$\mathbf{R}_x\mathbf{R}_x = \mathbf{I} \quad \text{and} \quad \langle \mathbf{R}_x U, \mathbf{R}_x V\rangle = \langle U, V\rangle\,, \quad \forall\, Z \in \mathbb{X}\,.$$

The properties (4.12) induce a symmetry in the governing equations. Act on (3.48) with $\mathbf{R}_x$

$$\mathbf{R}_x\mathbf{M}Z_t + \mathbf{R}_x\mathbf{K}Z_x + \mathbf{R}_x\mathbf{N}Z_y = \mathbf{R}_x\nabla S(Z)\,,$$

and use (4.12)

$$\mathbf{M}\mathbf{R}_x Z_t - \mathbf{K}\mathbf{R}_x Z_x + \mathbf{N}\mathbf{R}_x Z_y = \nabla S(\mathbf{R}_x Z)\,.$$

The implication is that

$$\mathbf{R}_x Z(-x, y, t) \quad \text{is a solution of (3.48) whenever } Z(x, y, t) \text{ is a solution}\,.$$

A similar argument generates a reversibility in the y-direction. Suppose there exists a reversor $\mathbf{R}_y$ acting on $\mathbb{X}$ satisfying

$$\begin{aligned}&\mathbf{R}_y\mathbf{M} = \mathbf{M}\mathbf{R}_y, \quad \mathbf{R}_y\mathbf{K} = \mathbf{K}\mathbf{R}_y, \quad \mathbf{R}_y\mathbf{N} = -\mathbf{N}\mathbf{R}_y, \quad \text{and}\\ &S(\mathbf{R}_y Z) = S(Z), \quad \forall\, Z \in \mathbb{X}.\end{aligned} \tag{4.13}$$

Act on (3.48) with $\mathbf{R}_y$

$$\mathbf{R}_y\mathbf{M}Z_t + \mathbf{R}_y\mathbf{K}Z_x + \mathbf{R}_y\mathbf{N}Z_y = \mathbf{R}_y\nabla S(Z)\,,$$

and use (4.13)

$$\mathbf{M}\mathbf{R}_y Z_t + \mathbf{K}\mathbf{R}_y Z_x - \mathbf{N}\mathbf{R}_y Z_y = \nabla S(\mathbf{R}_y Z)\,.$$

The implication is that

$$\mathbf{R}_y Z(x, -y, t) \quad \text{is a solution of (3.48) whenever } Z(x, y, t) \text{ is a solution}\,. \tag{4.14}$$

4.2.1 *Reversible Solitary Waves*

The solitary wave equation (4.7) can also be reversible in the ξ-direction. Equation (4.7) is reversible when there exists an involution $\mathbf{R}_\xi$ with the properties

$$\mathbf{R}_\xi\mathbf{J}(c, \ell) = -\mathbf{J}(c, \ell)\mathbf{R}_\xi \quad \text{and} \quad S(\mathbf{R}_\xi\widehat{Z}) = S(\widehat{Z})\,. \tag{4.15}$$

With this property, there exists a solution $\mathbf{R}_\xi\widehat{Z}(-\xi)$ whenever a solution $\widehat{Z}(\xi)$ exists. A special case is when *the solitary wave itself* is reversible. This latter case happens when $\widehat{Z}(\xi, c, \ell)$ satisfies

$$\mathbf{R}_\xi\widehat{Z}(-\xi, c, \ell) = \widehat{Z}(\xi, c, \ell)\,, \quad \forall\, \xi \in \mathbb{R}\,,\ (c, \ell) \in \mathscr{C}\,. \tag{4.16}$$

4.3 Symmetry of $I_1(c, \ell)$ and $I_2(c, \ell)$

In this section we show the implication for the functions $I_1(c, \ell)$ and $I_2(c, \ell)$ of a system that has a reversor in the y-direction. As shown in (4.14), the implication is that $\mathbf{R}_y Z(x, -y, t)$ is a solution whenever $Z(x, y, t)$ is a solution. Using only the abstract form of the equations, we can prove the following implication of y-direction reflection.

Proposition 4.1 *When the multisymplectic system* (4.1) *has a reflection symmetry in the y-direction, the functionals I_1 and I_2, evaluated on a (c,ℓ)-family of solitary waves, have the properties*

$$I_1(c,-\ell) = I_1(c,\ell) \quad \text{and} \quad I_2(c,-\ell) = -I_2(c,\ell)\,. \tag{4.17}$$

Proof Start with the definitions,

$$I_1(c,\ell) = \int_{-\infty}^{+\infty} \frac{1}{2}\langle \mathbf{M}\widehat{Z}_\xi, \widehat{Z}\rangle\, d\xi \quad \text{and} \quad I_2(c,\ell) = \int_{-\infty}^{+\infty} \frac{1}{2}\langle \mathbf{N}\widehat{Z}_\xi, \widehat{Z}\rangle\, d\xi\,, \tag{4.18}$$

with $\widehat{Z}(\xi,c,l)$ satisfying $\mathbf{J}(c,\ell)\widehat{Z}_\xi = \nabla S(\widehat{Z})$, or

$$\mathbf{K}\widehat{Z}_\xi + c\mathbf{M}\widehat{Z}_\xi + \ell\mathbf{N}\widehat{Z}_\xi = \nabla S(\widehat{Z})\,. \tag{4.19}$$

Act on this equation with $\mathbf{R}_y$ (to simplify notation drop the subscript y in the remainder of the proof) and use (4.13),

$$\mathbf{K}(\mathbf{R}\widehat{Z})_\xi + c\mathbf{M}(\mathbf{R}\widehat{Z})_\xi - \ell\mathbf{N}(\mathbf{R}\widehat{Z})_\xi = \nabla S(\mathbf{R}\widehat{Z})\,.$$

For each (c,ℓ), suppose $\widehat{Z}(\xi,c,\ell)$ is a unique reversible solution of (4.19). It then follows that

$$\mathbf{R}\widehat{Z}(\xi,c,\ell) = \widehat{Z}(\xi,c,-\ell)\,. \tag{4.20}$$

Now, use this identity in (4.18),

$$\begin{aligned}
I_1(c,-\ell) &= \int_{-\infty}^{+\infty} \frac{1}{2}\langle \mathbf{M}\frac{d}{d\xi}\widehat{Z}(\xi,c,-\ell), \widehat{Z}(\xi,c,-\ell)\rangle\, d\xi \\
&= \int_{-\infty}^{+\infty} \frac{1}{2}\langle \mathbf{M}\frac{d}{d\xi}\mathbf{R}\widehat{Z}(\xi,c,\ell), \mathbf{R}\widehat{Z}(\xi,c,\ell)\rangle\, d\xi \\
&= \int_{-\infty}^{+\infty} \frac{1}{2}\langle \mathbf{RM}\frac{d}{d\xi}\widehat{Z}(\xi,c,\ell), \mathbf{R}\widehat{Z}(\xi,c,\ell)\rangle\, d\xi \\
&= \int_{-\infty}^{+\infty} \frac{1}{2}\langle \mathbf{M}\widehat{Z}_\xi, \widehat{Z}\rangle\, d\xi = I_1(c,\ell)\,.
\end{aligned}$$

Similarly,

$$\begin{aligned}
I_2(c,-\ell) &= \int_{-\infty}^{+\infty} \frac{1}{2}\langle \mathbf{N}\frac{d}{d\xi}\widehat{Z}(\xi,c,-\ell), \widehat{Z}(\xi,c,-\ell)\rangle\, d\xi \\
&= \int_{-\infty}^{+\infty} \frac{1}{2}\langle \mathbf{N}\frac{d}{d\xi}\mathbf{R}\widehat{Z}(\xi,c,\ell), \mathbf{R}\widehat{Z}(\xi,c,\ell)\rangle\, d\xi
\end{aligned}$$

$$= -\int_{-\infty}^{+\infty} \frac{1}{2} \langle \mathbf{RN} \frac{d}{d\xi} \widehat{Z}(\xi, c, \ell), \mathbf{R}\widehat{Z}(\xi, c, \ell) \rangle \, d\xi$$

$$= -\int_{-\infty}^{+\infty} \frac{1}{2} \langle \mathbf{N}\widehat{Z}_\xi, \widehat{Z} \rangle \, d\xi = -I_2(c, \ell) \,.$$

This completes the proof of (4.17). ■

Corollary 4.2 *A consequence of the above symmetry result is that*

$$\lim_{\ell \to 0} \det \begin{bmatrix} (I_1)_c & (I_1)_\ell \\ (I_2)_c & (I_2)_\ell \end{bmatrix} = (I_1)_c(c, 0)\, \widehat{B}(c) \quad where \quad \widehat{B}(c) = \lim_{\ell \to 0} \frac{1}{\ell} I_2(c, \ell) \,. \tag{4.21}$$

The functions $I_1(c, 0)$ and $\widehat{B}(c)$ can be determined from the y-independent state. In other words, when the system has a reflection symmetry in the transverse direction, the sign of the determinant in (4.11) can be determined from properties of the ℓ-independent state.

4.4 Example: the Kadomtsev–Petviashvili Equation

The KP solution can be used to give an explicit example of the evaluation of the functions $I_1(c, \ell)$ and $I_2(c, \ell)$. First, substitute the form of the solitary wave $\widehat{Z}(\xi, c, \ell)$ with $\xi = x + ct + \ell y + \xi_0$ into the $(\mathbf{q}, \mathbf{p})$ coordinates in (3.56)–(3.58),

$$\begin{aligned} \widehat{p}_1 &= \tfrac{d}{d\xi} \left(-\tfrac{1}{2} c \widehat{u} - \widehat{p}_2 - \sigma \ell \widehat{p}_4 \right) , \\ \widehat{p}_2 &= f(\widehat{u}) + \tfrac{d}{d\xi} \left(+\tfrac{1}{2} c \widehat{q}_2 + \widehat{q}_3 - \ell \widehat{q}_4 \right) , \\ \widehat{p}_3 &= \widehat{u} \,, \\ \widehat{p}_4 &= \tfrac{d}{d\xi} \left(\ell \widehat{q}_2 + \tfrac{1}{\sigma} \widehat{q}_4 \right) , \end{aligned} \tag{4.22}$$

with $\widehat{p}_3$ replaced by $\widehat{u}$, and

$$\left(\widehat{q}_1\right)_{\xi\xi} = \left(\widehat{q}_2\right)_\xi = \widehat{u} \,, \quad \text{and} \quad \widehat{q}_3 = \widehat{u}_\xi \,, \tag{4.23}$$

and

$$\frac{d}{d\xi} (\widehat{p}_4 - \ell \widehat{u}) = 0 \quad \text{and} \quad \frac{d}{d\xi} (\widehat{p}_1) = 0 \,. \tag{4.24}$$

Integrating these two equations gives $\widehat{p}_1 = A$ and $\widehat{p}_4 - \ell\widehat{u} = B$, where A, B are integration constants. Integrating the four equations in (4.22) gives

$$
\begin{aligned}
\widehat{q}_4 &= \sigma B\,,\\
\widehat{p}_2 &= f(\widehat{u}) + \tfrac{1}{2}c\widehat{u} + \widehat{u}_{\xi\xi} - \sigma\ell B\,,\\
\widehat{p}_3 &= \widehat{u}\,,\\
\widehat{p}_4 &= \ell\widehat{u} + B\,.
\end{aligned}
$$

Substituting $\widehat{p}_1 = A$ into the first of (4.22) and integrating gives

$$
A\xi + D = -\frac{1}{2}c\widehat{u} - \widehat{p}_2 - \sigma\ell\widehat{p}_4\,,
$$

where D is another constant of integration. Substitute for $\widehat{p}_2$ and $\widehat{p}_4$, and use $p_3 = \widehat{u}$,

$$
A\xi + D = -c\widehat{u} - f(\widehat{u}) - \widehat{u}_{\xi\xi} + \sigma\ell B - \sigma\ell(\ell\widehat{u} + B)\,.
$$

Setting $A = 0$ for bounded solutions, and setting $D = 0$ since $\widehat{u} \to 0$ as $\xi \to \pm\infty$, gives

$$
\widehat{u}_{\xi\xi} + (c + \sigma\ell^2)\widehat{u} + f(\widehat{u}) = 0\,,
$$

which has an explicit exact solution when

$$
f(u) = u^{p+1}/(p+1)\,.
$$

However, we can make significant progress in evaluating the properties of I_1 and I_2 without the exact solution.

Now substitute the solitary wave $(\widehat{\mathbf{q}}(\xi), \widehat{\mathbf{p}}(\xi))$ into the functionals $I_1(c, \ell)$ and $I_2(c, \ell)$. For I_1 we compute

$$
\begin{aligned}
I_1(c, \ell) &= \int_{-\infty}^{+\infty} \frac{1}{2}\langle \mathbf{M}\widehat{Z}_\xi, \widehat{Z}\rangle\, d\xi\\
&= \frac{1}{2}\int_{-\infty}^{+\infty} \left(\widehat{q}_2 \frac{d}{d\xi}\widehat{p}_3 - \widehat{p}_3 \frac{d}{d\xi}\widehat{q}_2\right) d\xi\\
&= -\int_{-\infty}^{+\infty} \widehat{p}_3 \frac{d}{d\xi}\widehat{q}_2\, d\xi\,.
\end{aligned}
$$

The last expression follows from integration by parts using exponential decay as $\xi \to \pm\infty$. Now, noting that $\widehat{p}_3 = \widehat{u}$ and $\frac{d}{d\xi}\widehat{q}_2 = \widehat{p}_3 = \widehat{u}$, the functional reduces to

$$
I_1(c, \ell) = -\int_{-\infty}^{+\infty} \widehat{u}^2\, d\xi\,.
$$

This functional is an even function of ℓ, confirming the property in the first part of (4.17). A similar argument leads to an expression for I_2,

$$I_2(c,\ell) = \int_{-\infty}^{+\infty} \frac{1}{2}\langle \mathbf{N}\widehat{Z}_\xi, \widehat{Z}\rangle \, d\xi$$
$$= \frac{1}{2}\int_{-\infty}^{+\infty} \left(\sigma(\widehat{p}_4 \frac{d}{d\xi}\widehat{q}_2 - \widehat{q}_2 \frac{d}{d\xi}\widehat{p}_4) + \widehat{q}_4 \frac{d}{d\xi}\widehat{p}_3 - \widehat{p}_3 \frac{d}{d\xi}\widehat{q}_4 \right) d\xi \, .$$

The solitary wave state has the coordinates: $\widehat{q}_4 = 0$ and $\widehat{p}_4 = \ell \widehat{p}_3 = \ell \widehat{u}$. Therefore, after an integration by parts,

$$I_2(c,\ell) = \sigma\ell \int_{-\infty}^{+\infty} \widehat{u}^2 \, d\xi \, .$$

$I_2(c,\ell)$ is clearly an odd function of ℓ agreeing with the prediction in the second part of (4.17). The expressions for $I_1(c,\ell)$ and $I_2(c,\ell)$ are the same for any nonlinearity $f(u)$ in (3.53) for which a solitary wave solution exists.

We finish this section by giving the calculation of the non-degeneracy condition (4.11) for finite ℓ, when $f(u) = u^{p+1}/(p+1)$, using the explicit solution

$$\widehat{u} = a(c,\ell)\text{sech}^{2/p}(\gamma(c,\ell)\xi) \, , \quad \xi = x + ct + \ell y + \xi_0 \, ,$$

with

$$\gamma^2 = \frac{p^2}{4} b^2 \, , \quad a^p = \frac{1}{2}(p+1)(p+2)b(c,\ell)^2 \, , \quad b^2 = -c - \sigma\ell^2 > 0 \, ,$$

with the condition $c + \sigma\ell^2 < 0$. Substituting $\widehat{u}$ into I_1 gives

$$I_1(c,\ell) = -\int_{-\infty}^{+\infty} \widehat{u}^2 \, d\xi = -a^2 \int_{-\infty}^{+\infty} \text{sech}^{4/p}(\gamma\xi) d\xi \, .$$

Define

$$m(k) = \int_{-\infty}^{+\infty} \text{sech}^{4/p}(z) \, \mathrm{d}z \, ,$$

which is independent of c and ℓ, where $z = \gamma\xi$. Then

$$I_1(c,\ell) = -\frac{a^2}{\gamma} m(p) \, .$$

Similarly,

$$I_2(c, \ell) = \frac{1}{\gamma} a^2 \sigma \ell m(p) .$$

Hence

$$\begin{aligned}\det \begin{bmatrix} (I_1)_c & (I_1)_\ell \\ (I_2)_c & (I_2)_\ell \end{bmatrix} &= \det \begin{bmatrix} (I_1)_c & (I_1)_\ell \\ -\sigma\ell(I_1)_c & -\sigma(I_1)_c - \sigma\ell(I_1)_\ell \end{bmatrix} \\ &= -\sigma(I_1)(I_1)_c = \sigma|I_1|(I_1)_c ,\end{aligned}$$

since $I_1 < 0$. We will see later, once we construct the Evans function, and deduce an instability condition, a solitary wave is unstable if the above determinant is positive. From that theory, a solitary wave of KP of the above form, with $c + \sigma\ell^2 < 0$, is transversely unstable if

$$\sigma \partial_c I_1(c, \ell) > 0 . \tag{4.25}$$

From the above formula, we find

$$(I_1)_c = \frac{a^2 m(p)}{p\gamma b^2}(p-4) \quad \text{sign}\Big((I_1)_c\Big) = \text{sign}(p-4) .$$

In other words the transverse instability for KP is independent of the angle between the direction of travel of the basic state and the x-axis.

We can also see from the explicit calculation that the formula (4.21) is operational here, since

$$(I_1)_c(c, 0) = \frac{m(p)}{\sqrt{c}p^2} [(p+1)(p+2)]^{2/p} ,$$

and

$$\widehat{B}(c) = \lim_{\ell\to 0} \frac{1}{\ell} I_2(c, \ell) = \frac{1}{\gamma} a^2 \sigma m(p) > 0 ,$$

confirming that the instability condition for $\ell = 0$ is the same as $\ell \neq 0$. This calculation recovers the well-known result for transverse instability of the generalised KdV solitary wave (e.g. Kadomtsev and Petviashvili [85], Ostrovskii and Shrira [113], Alexander et al. [5], and Allen and Rowlands [6]). A rigorous proof of spectral instability will be a consequence of the theory developed later in Chap. 7.

Chapter 5
Linearisation About Solitary Waves

The linearisation about a family of solitary waves, starting with (4.7), is

$$\mathbf{J}(c,\ell)\widehat{Z}_\xi = D^2S(\widehat{Z})Z\,. \tag{5.1}$$

This equation can be re-cast in terms of the symmetric linear operator $\mathbf{L}$, defined by

$$\mathbf{L}(\xi,c,\ell) := \mathbf{B}(\xi,c,\ell) - \mathbf{K}\partial_\xi - c\mathbf{M}\partial_\xi - \ell\mathbf{N}\partial_\xi\,. \tag{5.2}$$

The equivalence with (5.1) follows from

$$\mathbf{B}(\xi,c,\ell) - \mathbf{J}(c,\ell)\partial_\xi = \mathbf{L}(\xi,c,\ell)$$

and

$$\mathbf{B}(\xi,c,\ell) := D^2S(\widehat{Z})\,. \tag{5.3}$$

The linear operator $\mathbf{L}$ can also be interpreted as the second variation of the Lagrange functional in the constrained variational principle (4.10),

$$\mathbf{L}(\xi,c,\ell) = D^2H(\widehat{Z}) - cD^2I_1(\widehat{Z}) - \ell D^2I_2(\widehat{Z})\,. \tag{5.4}$$

The functional analysis of symmetric linear operators of this type is well established, and so we will be brief. Effectively, we will be working in a vector-valued $L^2(\mathbb{R})$ space, which we will denote by

$$\mathbb{H}(\mathbb{R}) := \mathbb{R}^8 \otimes L^2(\mathbb{R})\,,$$

and the domain of $\mathbf{L}$ is an appropriate subset of $\mathbb{H}(\mathbb{R})$. For our purposes, the important property is that functions $F : \mathbb{R}\times\mathscr{C} \to \mathbb{R}^8$ in the domain of $\mathbf{L}$ are

T. J. Burchell, T. J. Bridges, *Transverse Instability of Solitary Waves*, Mathematical Physics Studies, https://doi.org/10.1007/978-981-95-2046-6_5

smooth enough functions of ξ, c and ℓ (see hypothesis below). The tangent vector to the solitary wave, which is also the derivative of (4.5) with respect to the phase shift ξ_0, is in the kernel of $\mathbf{L}$,

$$\mathbf{L}\widehat{Z}_\xi = 0\,. \tag{5.5}$$

It is assumed that

$$\mathrm{Ker}(\mathbf{L}) \cap \mathbb{H}(\mathbb{R}) = \mathrm{span}\{\widehat{Z}_\xi\}\,. \tag{5.6}$$

In other words, the kernel is not larger. This requirement constitutes hypothesis (**H3**).

Also, by differentiating (4.7) with respect to c, ℓ,

$$\begin{aligned} \big(\mathbf{K} + c\mathbf{M} + \ell\mathbf{N}\big)(\widehat{Z}_c)_\xi + \mathbf{M}\widehat{Z}_\xi &= \mathbf{B}(\xi, c, \ell)\widehat{Z}_c\,, \\ \big(\mathbf{K} + c\mathbf{M} + \ell\mathbf{N}\big)(\widehat{Z}_\ell)_\xi + \mathbf{N}\widehat{Z}_\xi &= \mathbf{B}(\xi, c, \ell)\widehat{Z}_\ell\,, \end{aligned}$$

we find that

$$\mathbf{L}\widehat{Z}_c = \mathbf{M}\widehat{Z}_\xi \quad \text{and} \quad \mathbf{L}\widehat{Z}_\ell = \mathbf{N}\widehat{Z}_\xi\,. \tag{5.7}$$

These equations for $\widehat{Z}_c$ and $\widehat{Z}_\ell$ will appear in the analysis of the Evans function.

5.1 The Eigenfunctions in the Linearisation

The linearisation (5.1) is a linear Hamiltonian system since the matrix $\mathbf{J}(c, \ell)$ is skew-symmetric and invertible for all $(c, \ell) \in \mathscr{C}$, and $\mathbf{B}(\xi, c, \ell)$ is symmetric for all $(\xi, c, \ell) \in \mathbb{R} \times \mathscr{C}$.

The ξ-dependence of $\mathbf{B}$ is due to the solitary wave, and with the asymptotic condition (4.6) the operator $\mathbf{B}$ is asymptotic to a constant matrix

$$\lim_{\xi \to \pm\infty} \mathbf{B}(\xi, c, \ell) = \mathbf{B}^\infty(c, \ell)\,.$$

The "system at infinity" for the steady problem is

$$\mathbf{J}(c, \ell)Z_\xi = \mathbf{B}^\infty(c, \ell)Z\,. \tag{5.8}$$

It is a linear constant coefficient ODE and can be solved explicitly,

$$Z(\xi) = \sum_{j=1}^{8} q_j \zeta_j e^{\mu_j \xi}\,, \tag{5.9}$$

where q_j are arbitrary complex constants, and $\mu_j(0, 0, c, \ell)$ are the eigenvalues determined from

$$\Delta(\mu, 0, 0; c, \ell) := \det\left[\mathbf{B}^\infty(c, \ell) - \mu \mathbf{J}(c, \ell)\right] = 0\,.$$

The two zeros in the arguments of $\Delta(\mu, 0, 0; c, \ell)$ anticipate the introduction of two exponents λ, ν in Chap. 6. The vectors ζ_j in (5.9) are the eigenvectors satisfying

$$\left[\mathbf{B}^\infty(c, \ell) - \mu_j(0, 0, c, \ell)\mathbf{J}(c, \ell)\right]\zeta_j(0, 0, c, \ell) = 0\,, \quad j = 1, \ldots, 8\,. \tag{5.10}$$

It is assumed that these eigenvalues are simple (see hypothesis (**H4**) below). The adjoint of the eigenvalue equation is

$$\left[\mathbf{J}(c, \ell)^{-1}\mathbf{B}^\infty(c, \ell)\right]^* \Xi_j = \overline{\mu}_j \Xi_j\,, \quad \Xi_j \in \mathbb{C}^8\,. \tag{5.11}$$

However, by taking $\Xi_j = \overline{\mathbf{J}(c, \ell)\eta_j(0, 0, c, \ell)}$, the equation satisfied by $\eta_j(0, 0, c, \ell)$ is

$$\left[\mathbf{B}^\infty(c, \ell) + \mu_j(0, 0, c, \ell)\mathbf{J}(c, \ell)\right]\eta_j(0, 0, c, \ell) = 0\,, \quad j = 1, \ldots, 8\,. \tag{5.12}$$

The adjoint eigenvectors are normalised as

$$\langle \mathbf{J}(c, \ell)\eta_i(0, 0, c, \ell), \zeta_j(0, 0, c, \ell)\rangle = \mathbf{\Omega}(\eta_i, \zeta_j) = \delta_{i,j}\,, \quad i, j = 1, \ldots, 8\,. \tag{5.13}$$

The spectrum of $\mathbf{J}(c, \ell)^{-1}\mathbf{B}^\infty(c, \ell)$ is symmetric about the origin due to the symmetry of $\mathbf{B}^\infty(c, \ell)$ and the skew-symmetry of $\mathbf{J}(c, \ell)$.

Assume that all 8 eigenvalues are hyperbolic. Then Hamiltonian symmetry gives that the spectrum has a 4–4 splitting for all $(c, \ell) \in \mathscr{C}$. The eigenvalues in the splitting are numbered by

$$\mu_1 < \mu_2 < \mu_3 < \mu_4 < 0 < \mu_5 < \mu_6 < \mu_7 < \mu_8\,,$$

where

$$\mu_1 = -\mu_8\,, \quad \mu_2 = -\mu_7\,, \quad \mu_3 = -\mu_6\,, \quad \mu_4 = -\mu_5\,. \tag{5.14}$$

Together with the assumptions of real and simple eigenvalues, this forms hypothesis (**H4**).

Now go back to the ξ-dependent linear system (5.1) and let $E^u(E^s)$ be the ξ-dependent unstable (stable) subspaces of (5.8). Introduce vector-valued functions

$$\mathbf{a}^\pm(\xi)\,, \quad \mathbf{b}^\pm(\xi)\,, \quad \text{and} \quad \mathbf{c}^\pm(\xi)\,,$$

to represent the stable and unstable spaces; that is,

$$E^s(\xi, 0, 0) = \text{span}\{\widehat{Z}_\xi, \mathbf{a}^+, \mathbf{b}^+, \mathbf{c}^+\} \quad \text{and} \quad E^u(\xi, 0, 0) = \text{span}\{\widehat{Z}_\xi, \mathbf{a}^-, \mathbf{b}^-, \mathbf{c}^-\}, \tag{5.15}$$

where $\widehat{Z}_\xi$ is the tangent vector to the solitary wave which decays exponentially as $\xi \to \pm\infty$, and $\mathbf{a}^\pm(\xi), \mathbf{b}^\pm(\xi), \mathbf{c}^\pm(\xi)$ are solutions of the linearised system (5.1) which decay exponentially as $\xi \to +\infty$ (E^s) or decay exponentially as $\xi \to -\infty$ (E^u). The precise statement is below.

Without loss of generality, label μ_4 and μ_5 as the decay rates of $\widehat{Z}_\xi$ as $\xi \to \pm\infty$. The precise asymptotics of $\widehat{Z}_\xi$ are discussed below in Sect. 5.3 when normalisation of the volume form is presented. The other six eigenvalues give the decay rates of $\mathbf{a}^\pm(\xi), \mathbf{b}^\pm(\xi), \mathbf{c}^\pm(\xi)$. They are labelled as follows:

$$\begin{array}{ll}
\lim_{\xi\to+\infty} e^{-\mu_3\xi}\mathbf{a}^+ = \zeta_3\,, & \lim_{\xi\to-\infty} e^{-\mu_6\xi}\mathbf{a}^- = \zeta_6\,,\\
\lim_{\xi\to+\infty} e^{-\mu_2\xi}\mathbf{b}^+ = \zeta_2\,, & \lim_{\xi\to-\infty} e^{-\mu_7\xi}\mathbf{b}^- = \zeta_7\,,\\
\lim_{\xi\to+\infty} e^{-\mu_1\xi}\mathbf{c}^+ = \zeta_1\,, & \lim_{\xi\to-\infty} e^{-\mu_8\xi}\mathbf{c}^- = \zeta_8\,,
\end{array}$$

with the convergence exponential. In general $\mathbf{a}^\pm$, $\mathbf{b}^\pm$, $\mathbf{c}^\pm$ are *not bounded* as $\xi \to \mp\infty$.

Definition 2 A subspace E of a $2n$-dimensional symplectic vector space is Lagrangian if it has dimension n and $\Omega(\xi_i, \xi_j) = 0$ for all $\xi_i, \xi_j \in E$.

Definition 3 For any linear subspace W of a symplectic vector space (V, Ω) the symplectic complement of W is defined as

$$W^\perp := \{v \in V \mid \Omega(v, w) = 0, \ \forall\, w \in W\}.$$

N.B. If V is finite-dimensional then

$$\dim W + \dim W^\perp = \dim V\,. \tag{5.16}$$

The subspace W is said to be Lagrangian if and only if $W = W^\perp$.

Proposition 5.1 *$E^s(\xi, 0, 0)$ and $E^u(\xi, 0, 0)$ are Lagrangian subspaces of $\mathbb{X} = \mathbb{R}^8$ with respect to the symplectic structure $\mathbf{\Omega}$ in* (4.3).

Proof Let Z_1, Z_2 be any two solutions of (5.1). Then a direct calculation using (5.1) and the skew-symmetry of $\mathbf{J}(c, \ell)$ gives

$$\frac{d}{d\xi}\mathbf{\Omega}(Z_1, Z_2) = 0\,.$$

This proves that $\mathbf{\Omega}(Z_1, Z_2)$ is a constant for all $\xi \in \mathbb{R}$. Now if $Z_1, Z_2 \in E^s(\xi, 0, 0)$ they both go to zero as $\xi \to \infty$. Likewise, if $Z_1, Z_2 \in E^u(\xi, 0, 0)$ they both go to zero as $\xi \to -\infty$. In both these cases we can conclude that

$$\boldsymbol{\Omega}(Z_1, Z_2) = 0 \quad \forall \xi \in \mathbb{R}, \tag{5.17}$$

hence $E^s \subseteq (E^s)^\perp$ and $E^u \subseteq (E^u)^\perp$. Now use (5.16) and $\dim E^s = \dim E^u = 4$ to conclude that $E^s = (E^s)^\perp$ and $E^u = (E^u)^\perp$. ■

The pairing $\boldsymbol{\Omega}(Z_1, Z_2)$ is not necessarily zero when $Z_1 \in E^u(\xi, 0, 0)$ and $Z_2 \in E^s(\xi, 0, 0)$ (or vice versa); however, we can conclude that:

Corollary 5.2 *For all* $Z^- \in \operatorname{span}\{\mathbf{a}^-, \mathbf{b}^-, \mathbf{c}^-\}$ *and* $Z^+ \in \operatorname{span}\{\mathbf{a}^+, \mathbf{b}^+, \mathbf{c}^+\}$, $\boldsymbol{\Omega}(Z^-, Z^+)$ *is independent of* ξ.

Remark This corollary ensures that the Lazutkin–Treschev invariant (defined below) is independent of ξ.

5.2 Transversality and the Lazutkin–Treschev Invariant

The solitary wave solution is said to be *transversely constructed* when the intersection between the stable and unstable manifolds is transverse. We will first give an intuitive condition for transversality, and then show that it can be transformed into an elegant symplectic formulation known as the Lazutkin–Treschev invariant.

The stable and unstable spaces, representing the tangent spaces to the stable and unstable manifolds, are given in (5.15). It is the normal space in each

$$E^{sn}(\xi, 0, 0) = \operatorname{span}\{\mathbf{a}^+, \mathbf{b}^+, \mathbf{c}^+\} \quad \text{and} \quad E^{un}(\xi, 0, 0) = \operatorname{span}\{\mathbf{a}^-, \mathbf{b}^-, \mathbf{c}^-\}, \tag{5.18}$$

that are used to define transversality. An intuitive definition is

$$\mathbf{a}^- \wedge \mathbf{b}^- \wedge \mathbf{c}^- \wedge \mathbf{a}^+ \wedge \mathbf{b}^+ \wedge \mathbf{c}^+ \neq 0, \quad \forall \xi \in \mathbb{R}. \tag{5.19}$$

Now we appeal to the formula in Proposition 5.4 of [40]

$$\mathbf{a}^- \wedge \mathbf{b}^- \wedge \mathbf{c}^- \wedge \mathbf{a}^+ \wedge \mathbf{b}^+ \wedge \mathbf{c}^+ = \det \begin{bmatrix} \boldsymbol{\Omega}(\mathbf{a}^-, \mathbf{a}^+) & \boldsymbol{\Omega}(\mathbf{b}^-, \mathbf{a}^+) & \boldsymbol{\Omega}(\mathbf{c}^-, \mathbf{a}^+) \\ \boldsymbol{\Omega}(\mathbf{a}^-, \mathbf{b}^+) & \boldsymbol{\Omega}(\mathbf{b}^-, \mathbf{b}^+) & \boldsymbol{\Omega}(\mathbf{c}^-, \mathbf{b}^+) \\ \boldsymbol{\Omega}(\mathbf{a}^-, \mathbf{c}^+) & \boldsymbol{\Omega}(\mathbf{b}^-, \mathbf{c}^+) & \boldsymbol{\Omega}(\mathbf{c}^-, \mathbf{c}^+) \end{bmatrix} \varpi + \Upsilon \varpi,$$

where ϖ is a volume form on the 6-dimensional ambient space, and

$$\begin{aligned}\Upsilon = -\boldsymbol{\Omega}(\mathbf{a}^-,\mathbf{b}^-)&\left[\boldsymbol{\Omega}(\mathbf{c}^-,\mathbf{a}^+)\boldsymbol{\Omega}(\mathbf{b}^+,\mathbf{c}^+) - \det\begin{pmatrix}\boldsymbol{\Omega}(\mathbf{c}^-,\mathbf{b}^+) & \boldsymbol{\Omega}(\mathbf{c}^-,\mathbf{c}^+)\\ \boldsymbol{\Omega}(\mathbf{a}^+,\mathbf{b}^+) & \boldsymbol{\Omega}(\mathbf{a}^+,\mathbf{c}^+)\end{pmatrix}\right]\\ +\boldsymbol{\Omega}(\mathbf{a}^-,\mathbf{c}^-)&\left[\boldsymbol{\Omega}(\mathbf{b}^-,\mathbf{a}^+)\boldsymbol{\Omega}(\mathbf{b}^+,\mathbf{c}^+) - \det\begin{pmatrix}\boldsymbol{\Omega}(\mathbf{b}^-,\mathbf{b}^+) & \boldsymbol{\Omega}(\mathbf{b}^-,\mathbf{c}^+)\\ \boldsymbol{\Omega}(\mathbf{a}^+,\mathbf{b}^+) & \boldsymbol{\Omega}(\mathbf{a}^+,\mathbf{c}^+)\end{pmatrix}\right]\\ -\boldsymbol{\Omega}(\mathbf{b}^-,\mathbf{c}^-)&\left[\boldsymbol{\Omega}(\mathbf{a}^-,\mathbf{a}^+)\boldsymbol{\Omega}(\mathbf{b}^+,\mathbf{c}^+) - \det\begin{pmatrix}\boldsymbol{\Omega}(\mathbf{a}^-,\mathbf{b}^+) & \boldsymbol{\Omega}(\mathbf{a}^-,\mathbf{c}^+)\\ \boldsymbol{\Omega}(\mathbf{a}^+,\mathbf{b}^+) & \boldsymbol{\Omega}(\mathbf{a}^+,\mathbf{c}^+)\end{pmatrix}\right].\end{aligned} \tag{5.20}$$

A proof is given in [40]. This formula is remarkable because it relates an elementary wedge product between vectors to a determinant with entries expressed in terms of a symplectic form. It is exact and independent of the choice of symplectic form. Here the symplectic form is taken to be the one based on the symplectic operator $\mathbf{J}(c,\ell)$.

The extra term Υ makes the formula less elegant and unwieldy. However, we can use the property in Proposition 5.1 that the stable and unstable subspaces are Lagrangian to nullify Υ. Proposition 5.1 gives

$$\begin{aligned}&\boldsymbol{\Omega}(\mathbf{a}^-,\mathbf{b}^-) = \boldsymbol{\Omega}(\mathbf{b}^-,\mathbf{c}^-) = \boldsymbol{\Omega}(\mathbf{a}^-,\mathbf{c}^-) = 0 \quad \text{and}\\ &\boldsymbol{\Omega}(\mathbf{a}^+,\mathbf{b}^+) = \boldsymbol{\Omega}(\mathbf{b}^+,\mathbf{c}^+) = \boldsymbol{\Omega}(\mathbf{a}^+,\mathbf{c}^+) = 0\,.\end{aligned}$$

Substituting these into Υ gives $\Upsilon = 0$, in which case the formula (5.20) reduces to

$$\mathbf{a}^- \wedge \mathbf{b}^- \wedge \mathbf{c}^- \wedge \mathbf{a}^+ \wedge \mathbf{b}^+ \wedge \mathbf{c}^+ = \Pi\varpi\,,$$

with

$$\Pi(c,\ell) = \det\begin{bmatrix}\boldsymbol{\Omega}(\mathbf{a}^-,\mathbf{a}^+) & \boldsymbol{\Omega}(\mathbf{b}^-,\mathbf{a}^+) & \boldsymbol{\Omega}(\mathbf{c}^-,\mathbf{a}^+)\\ \boldsymbol{\Omega}(\mathbf{a}^-,\mathbf{b}^+) & \boldsymbol{\Omega}(\mathbf{b}^-,\mathbf{b}^+) & \boldsymbol{\Omega}(\mathbf{c}^-,\mathbf{b}^+)\\ \boldsymbol{\Omega}(\mathbf{a}^-,\mathbf{c}^+) & \boldsymbol{\Omega}(\mathbf{b}^-,\mathbf{c}^+) & \boldsymbol{\Omega}(\mathbf{c}^-,\mathbf{c}^+)\end{bmatrix}. \tag{5.21}$$

This function is the Lazutkin–Treschev invariant. It was named in [40], where a new proof, leading to the formula above, was given. This invariant was first proposed in Treschev [135], who gives a different construction, and it generalises the Lazutkin invariant, which holds in $\mathbb{R}^4$ [40, 67, 68]. It is independent of ξ, invariant under $\mathbf{J}(c,\ell)$-symplectic changes of coordinates, and non-zero precisely when the homoclinic orbit is transversely constructed. To fix the sign of the Lazutkin–Treschev invariant an orientation is needed.

5.2.1 Orientation

Let $\{\mathbf{e}_1, \ldots, \mathbf{e}_8\}$ be any fixed (independent of c, ℓ) basis, not necessarily the standard basis, for the phase space, $\mathbb{R}^8$ and fix an orientation by defining

$$\varpi := \mathbf{e}_1 \wedge \mathbf{e}_2 \wedge \mathbf{e}_3 \wedge \mathbf{e}_4 \wedge \mathbf{e}_5 \wedge \mathbf{e}_6 \wedge \mathbf{e}_7 \wedge \mathbf{e}_8 \,. \tag{5.22}$$

The eigenvectors are related to the orientation by

$$\zeta_1(0,0,c,\ell) \wedge \ldots \wedge \zeta_8(0,0,c,\ell) = K(0,0,c,\ell)\varpi \,, \quad \text{with} \,, \quad \forall\,(c,\ell) \in \mathscr{C} \,, \tag{5.23}$$

for some non-zero function $K(0,0,c,\ell)$. To fix the sign of Π we have to synchronise the orientations of the stable and unstable subspace, relative to the ambient space. The unstable subspace satisfies

$$\operatorname{span}\{\widehat{Z}_\xi, \mathbf{a}^-, \mathbf{b}^-, \mathbf{c}^-\} \to \operatorname{span}\{\zeta_5, \zeta_6, \zeta_7, \zeta_8\} \quad \text{as} \quad \xi \to -\infty \,. \tag{5.24}$$

Hence there exists a constant C^- with the property that

$$\begin{aligned} &e^{-(\mu_5+\mu_6+\mu_7+\mu_8)\xi}\, \widehat{Z}_\xi(\xi) \wedge \mathbf{a}^-(\xi) \wedge \mathbf{b}^-(\xi) \wedge \mathbf{c}^-(\xi) \\ &\quad \to C^-\zeta_5 \wedge \zeta_6 \wedge \zeta_7 \wedge \zeta_8 \quad \text{as} \quad \xi \to -\infty \,. \end{aligned} \tag{5.25}$$

Similarly, the stable subspace satisfies

$$\operatorname{span}\{\widehat{Z}_\xi, \mathbf{a}^+, \mathbf{b}^+, \mathbf{c}^+\} \to \operatorname{span}\{\zeta_1, \zeta_2, \zeta_3, \zeta_4\} \quad \text{as} \quad \xi \to +\infty \,. \tag{5.26}$$

Hence there exists a constant C^+ with the property that

$$\begin{aligned} &e^{-(\mu_1+\mu_2+\mu_3+\mu_4)\xi}\, \widehat{Z}_\xi(\xi) \wedge \mathbf{a}^+(\xi) \wedge \mathbf{b}^+(\xi) \wedge \mathbf{c}^+(\xi) \\ &\quad \to C^+\zeta_1 \wedge \zeta_2 \wedge \zeta_3 \wedge \zeta_4 \quad \text{as} \quad \xi \to +\infty \,. \end{aligned} \tag{5.27}$$

Take ξ to $-\xi$ in this formula, and use (5.14),

$$\begin{aligned} &e^{-(\mu_5+\mu_6+\mu_7+\mu_8)\xi}\, \widehat{Z}_\xi(-\xi) \wedge \mathbf{a}^+(-\xi) \wedge \mathbf{b}^+(-\xi) \wedge \mathbf{c}^+(-\xi) \\ &\quad \to C^+\zeta_1 \wedge \zeta_2 \wedge \zeta_3 \wedge \zeta_4 \quad \text{as} \quad \xi \to -\infty \,. \end{aligned} \tag{5.28}$$

Take the wedge product between (5.25) and (5.28) and take the limit $\xi \to -\infty$,

$$\begin{aligned} &e^{-2(\mu_5+\mu_6+\mu_7+\mu_8)\xi}\, \widehat{Z}_\xi(-\xi) \wedge \mathbf{a}^+(-\xi) \wedge \mathbf{b}^+(-\xi) \wedge \mathbf{c}^+(-\xi) \wedge \widehat{Z}_\xi(\xi) \wedge \mathbf{a}^-(\xi) \\ &\quad \wedge \mathbf{b}^-(\xi) \wedge \mathbf{c}^-(\xi) \\ &\longrightarrow C^+C^-\zeta_1 \wedge \zeta_2 \wedge \zeta_3 \wedge \zeta_4 \wedge \zeta_5 \wedge \zeta_6 \wedge \zeta_7 \wedge \zeta_8 \,, \end{aligned}$$

or

$$\begin{aligned}\lim_{\xi\to-\infty} e^{-2(\mu_5+\mu_6+\mu_7+\mu_8)\xi}\, \widehat{Z}_\xi(-\xi)\wedge \mathbf{a}^+(-\xi)\wedge \mathbf{b}^+(-\xi)\wedge \mathbf{c}^+(-\xi)\wedge \widehat{Z}_\xi(\xi)\\ \wedge\mathbf{a}^-(\xi)\wedge \mathbf{b}^-(\xi)\wedge \mathbf{c}^-(\xi)\\ = C^+C^-K(0,0,c,\ell)\varpi\,,\end{aligned}$$

using (5.23). Taking the sign of $C^+C^-K(0,0,c,\ell)$ to be fixed throughout the analysis means that it is not possible to change the sign of any of $\{\mathbf{a}^\pm, \mathbf{b}^\pm, \mathbf{c}^\pm\}$ individually, they can only have their signs changed in pairs. By considering how a paired change of signs affects the matrix in (5.21) it is clear that this fixes the sign of Π. These properties of $\Pi(c,\ell)$ and the choice of orientation constitute hypothesis **(H5)**.

5.3 Asymptotics of $\widehat{Z}_\xi$ and the Factor χ

In this subsection, we establish the way that $\widehat{Z}_\xi$ approaches the eigenvectors at infinity. Firstly, by the choice of eigenvectors (5.1), there exists real numbers $\chi_\pm$ with

$$\lim_{\xi\to-\infty} e^{-\mu_5\xi}\widehat{Z}_\xi = \chi^-\zeta_5 \quad\text{and}\quad \lim_{\xi\to+\infty} e^{+\mu_5\xi}\widehat{Z}_\xi = \chi^+\eta_5\,. \tag{5.29}$$

These expressions can be inverted to give formulae for $\chi^\pm$,

$$\chi^- = \lim_{\xi\to-\infty} e^{-\mu_5\xi}\,\mathbf{\Omega}(\eta_5, \widehat{Z}_\xi) \quad\text{and}\quad \chi^+ = \lim_{\xi\to+\infty} e^{+\mu_5\xi}\,\mathbf{\Omega}(\widehat{Z}_\xi, \zeta_5)\,. \tag{5.30}$$

Define

$$\chi = (\chi^+\chi^-)^{-1}\,. \tag{5.31}$$

This factor χ will appear in the construction of the Evans function. The sign of χ is affected by reversibility of the solitary wave.

Proposition 5.3 *When the basic state is reversible,*

$$\mathrm{sign}(\chi) = \mathrm{sign}\,[\mathbf{\Omega}(\zeta_5, \mathbf{R}\zeta_5)],$$

where $\mathbf{R}$ *is the reversor in* (4.16) *with the subscript* ξ *dropped for brevity.*

Proof Differentiating (4.16) with respect to ξ gives

$$-\mathbf{R}\widehat{Z}_\xi(-\xi, c, \ell) = \widehat{Z}_\xi(\xi, c, \ell)\,. \tag{5.32}$$

Now use the formula (5.29) for the $\widehat{Z}_\xi$ asymptotics as $\xi \to -\infty$, take $\xi \to -\xi$ in this equation, and act on it with **R**,

$$\lim_{\xi \to +\infty} e^{+\mu_5 \xi} \mathbf{R} \widehat{Z}_\xi(-\xi, c, \ell) = \chi^- \mathbf{R} \zeta_5 .$$

Substituting (5.32) then gives

$$\lim_{\xi \to +\infty} e^{+\mu_5 \xi} \widehat{Z}_\xi(\xi, c, \ell) = -\chi^- \mathbf{R} \zeta_5 . \tag{5.33}$$

Now use the second asymptotic formula in (5.29) for $\xi \to +\infty$

$$\lim_{\xi \to +\infty} e^{+\mu_5 \xi} \widehat{Z}_\xi(\xi, c, \ell) = \chi^+ \eta_5 . \tag{5.34}$$

Combining (5.33) and (5.34),

$$-\chi^- \mathbf{R} \zeta_5 = \chi^+ \eta_5 . \tag{5.35}$$

Taking the pairing with $\mathbf{J}\zeta_5$ on both sides of (5.35) then results in

$$\begin{aligned} -\chi^- \langle \mathbf{J}\zeta_5, \mathbf{R}\zeta_5 \rangle &= \chi^+ \langle \mathbf{J}\zeta_5, \eta_5 \rangle , \\ \Rightarrow \quad \chi^- \mathbf{\Omega}(\zeta_5, \mathbf{R}\zeta_5) &= \chi^+ \mathbf{\Omega}(\eta_5, \zeta_5) , \\ \Rightarrow (\chi^-)^2 \mathbf{\Omega}(\zeta_5, \mathbf{R}\zeta_5) &= \chi^{-1} . \end{aligned} \tag{5.36}$$

Considering the signs of each side of this expression then proves the proposition. ■

Chapter 6
Spectral Stability and the Evans Function

In this chapter we study the linearisation of the full time-dependent PDE (4.1) about the solitary wave solution (4.5). Substitute $\widehat{Z}(\xi) + \varepsilon Z(\xi, y, t)$ into (4.1),

$$\mathbf{M}(c\widehat{Z}_\xi + \varepsilon Z_t + \varepsilon c Z_\xi) + \mathbf{K}(\widehat{Z}_\xi + \varepsilon Z_\xi) + \mathbf{N}(\ell\widehat{Z}_\xi + \varepsilon Z_y + \varepsilon\ell Z_\xi) = \nabla S(\widehat{Z} + \varepsilon Z) .$$

Rearrange,

$$\mathbf{J}(c, \ell)\widehat{Z}_\xi + \varepsilon(\mathbf{M}Z_t + \mathbf{J}(c, \ell)Z_\xi + \mathbf{N}Z_y) = \nabla S(\widehat{Z}) + \varepsilon D^2 S(\widehat{Z})Z + \mathcal{O}(\varepsilon^2) ,$$

subtract off the equation for $\widehat{Z}$, and take the limit $\varepsilon \to 0$,

$$\mathbf{M}Z_t + \mathbf{J}(c, \ell)Z_\xi + \mathbf{N}Z_y = \mathbf{B}(\xi, c, \ell)Z, \quad Z \in \mathbb{X} , \tag{6.1}$$

where $\mathbf{B}$ is the Hessian of S, defined in (5.3). Introduce the spectral ansatz

$$Z(\xi, y, t) = e^{\lambda t + \mathrm{i}\nu y}\mathbf{u}(\xi, \lambda, \nu) , \tag{6.2}$$

with $\lambda \in \mathbb{C}$ and $\nu \in \mathbb{R}$. The $e^{\lambda t}$ part of this ansatz just introduces the spectral parameter λ. The ν part can be justified various ways. The linear PDE (6.1) has non-constant coefficients that are independent of y and so a Fourier transform can be taken in the y-direction. This would require a more formal approach to the construction. Here we just take the asserted form as an axiom (in the hydrodynamic stability literature; this is called a "normal mode ansatz" [63]).

With the spectral ansatz (6.2), the ν-dependent eigenvalue problem for λ is

$$\mathbf{u}_\xi = \mathbf{A}(\xi, \lambda, \nu)\mathbf{u}, \quad \mathbf{u} \in V , \quad (\lambda, \nu) \in \Lambda , \tag{6.3}$$

T. J. Burchell, T. J. Bridges, *Transverse Instability of Solitary Waves*, Mathematical Physics Studies, https://doi.org/10.1007/978-981-95-2046-6_6

for some open set $\Lambda \subset \mathbb{C} \times \mathbb{R}$, with

$$\mathbf{A}(\xi, \lambda, \nu) := \mathbf{J}(c, \ell)^{-1}[\mathbf{B}(\xi, c, \ell) - \lambda\mathbf{M} - \mathrm{i}\nu\mathbf{N}]. \tag{6.4}$$

Let V be the complexification of $\mathbb{X}$. It is an eight-dimensional complex vector space, with dual V^*. It is effectively $\mathbb{C}^8$, but the structure is clarified by using the more general vector space setting.

Definition 4 A mode $(\lambda, \mathbf{u})$, associated with the transverse exponent ν, is said to be spectrally unstable if there exists a solution of (6.3), for some $\nu \in \mathbb{R}$ and some $\lambda \in \mathbb{C}$ with $\mathrm{Re}(\lambda) > 0$, and

$$\mathbf{u}(\cdot, \lambda) \in \mathbb{H}^{\mathbb{C}}(\mathbb{R}) := V \otimes L^2(\mathbb{R}).$$

With the asymptotic condition (4.6) and analyticity of the eigenvalues, we have that

$$\int_{-\infty}^{+\infty} \|\mathbf{A}(\xi, \lambda, \nu) - \mathbf{A}^\infty(\lambda, \nu)\| \, \mathrm{d}\xi < +\infty, \quad (\lambda, \nu) \in \Lambda. \tag{6.5}$$

In this integral the matrix $\mathbf{A}$ at infinity is defined by

$$\mathbf{A}^\infty(\lambda, \nu) := \lim_{\xi \to \pm\infty} \mathbf{A}(\xi, \lambda, \nu) = \mathbf{J}^{-1}[\mathbf{B}^\infty - \lambda\mathbf{M} - \mathrm{i}\nu\mathbf{N}] \quad \forall (\lambda, \nu) \in \Lambda, \tag{6.6}$$

with the dependence on c, ℓ suppressed to simplify notation. The set Λ is defined based on the position of the eigenvalues of $\mathbf{A}^\infty(\lambda, \nu)$. The spectrum of $\mathbf{A}^\infty(\lambda, \nu)$ consists of eight eigenvalues

$$\begin{aligned}\sigma(\mathbf{A}^\infty(\lambda, \nu)) = \{&\mu_1(\lambda, \nu, c, \ell), \mu_2(\lambda, \nu, c, \ell), \mu_3(\lambda, \nu, c, \ell), \mu_4(\lambda, \nu, c, \ell),\\ &\mu_5(\lambda, \nu, c, \ell), \mu_6(\lambda, \nu, c, \ell), \mu_7(\lambda, \nu, c, \ell), \mu_8(\lambda, \nu, c, \ell)\},\end{aligned} \tag{6.7}$$

and the set Λ is defined as an open set in the space $\mathbb{C} \times \mathbb{R}$, including the origin, such that for all $(\lambda, \nu) \in \Lambda$ the eight eigenvalues of $\mathbf{A}^\infty(\lambda, \nu)$ satisfy

$$\begin{aligned}\mathrm{Re}(\mu_1) \le \mathrm{Re}(\mu_2) \le \mathrm{Re}(\mu_3) \le \mathrm{Re}(\mu_4) < 0 < \mathrm{Re}(\mu_5)\\ \le \mathrm{Re}(\mu_6) \le \mathrm{Re}(\mu_7) \le \mathrm{Re}(\mu_8).\end{aligned} \tag{6.8}$$

These eight spatial eigenvalues are assumed to be simple for all $(\lambda, \nu) \in \Lambda$.

The continuous spectrum is parameterised by (κ, ν) and defined by

$$\sigma^{\mathrm{cont}}(\kappa, \nu) = \left\{ \lambda \in \mathbb{C} \, : \, \det[\mathbf{B}^\infty - \lambda\mathbf{M} - \mathrm{i}\nu\mathbf{N} - \mathrm{i}\kappa\mathbf{J}] = 0, \quad \nu \in \mathbb{R}, \, \kappa \in \mathbb{R} \right\}. \tag{6.9}$$

We do not assume that the continuous spectrum is stable. However, when the continuous spectrum passes through the origin, we assume that the Evans function, constructed below, is twice differentiable there. This latter property can be ensured by various strategies: (a) using the gap lemma to shift the continuous spectrum (e.g. Chapter 10 of [86] and references therein), (b) using analytic continuation [31], and (c) expanding the μ-eigenvalues in Taylor series near $\lambda = 0$ [30].

The trace of $\mathbf{A}(\xi, \lambda, \nu)$ is independent of ξ and can be simplified to

$$\begin{aligned}\mathrm{Tr}(\mathbf{A}(\xi, \lambda, \nu)) &= \mathrm{Tr}\big(-\lambda\mathbf{J}(c, \ell)^{-1}\mathbf{M}\big) + \mathrm{Tr}\big(-\mathtt{i}\nu\mathbf{J}(c, \ell)^{-1}\mathbf{N}\big) \\ &:= \lambda\tau_1(c, \ell) + \mathtt{i}\nu\tau_2(c, \ell)\,,\end{aligned} \tag{6.10}$$

for some scalar-valued functions $\tau_1(c, \ell)$, $\tau_2(c, \ell)$. Explicit expressions for $\tau_1(c, \ell)$ and $\tau_2(c, \ell)$ can be obtained but are not needed.

The above properties of the eigenvalues and eigenvectors at infinity, the definition of the open set $\Lambda \subset \mathbb{C} \times \mathbb{R}$, the properties of the continuous spectrum near $\lambda = 0$, and the existence and uniformity of the integral condition (6.5) form hypothesis (**H6**).

6.1 Constructing the Evans Function

The Evans function is a complex-valued function, which will be denoted by $D(\lambda, \nu)$, with the property: if for some $\nu \in \mathbb{R}$ the function $D(\lambda, \nu) = 0$ for some $\lambda \in \mathbb{C}$, with $\mathrm{Re}(\lambda) > 0$, then λ corresponds to an unstable eigenvalue. It measures the intersection between the stable and unstable manifolds of the dynamical system (6.3). There is now a vast literature on the Evans function, particularly in $1 + 1$, and its definition, variants, history, and properties can be found in Kapitula and Promislow [86].

The Evans function will first be constructed as the wedge product of individual vector-valued solutions of (6.3), then an equivalent definition which pairs solutions of (6.3) with solutions of the adjoint equation will be derived. Associated with each of the simple eigenvalues (6.8) is an eigenvector

$$\big[\mathbf{A}^{\infty}(\lambda, \nu) - \mu_j(\lambda, \nu)\mathbf{I}\big]\zeta_j(\lambda, \nu) = 0, \quad \zeta_j : \Lambda \to V, \quad j = 1, \ldots, 8, \quad (\lambda, \nu) \in \Lambda. \tag{6.11}$$

Since $\mu_j(\lambda, \nu)$ are simple and therefore analytic functions of λ, the eigenvectors can also be constructed to be analytic functions of λ. Using standard asymptotic theory for ODEs, there are eight (ξ, λ, ν)-dependent vectors satisfying

$$(\mathbf{u}_j)_\xi = \mathbf{A}(\xi, \lambda, \nu)\mathbf{u}_j\,, \quad j = 1, \ldots, 8\,, \tag{6.12}$$

with the asymptotic properties

$$\lim_{\xi\to+\infty} e^{-\mu_1(\lambda,\nu)\xi}\mathbf{u}_1(\xi,\lambda,\nu) = \zeta_1(\lambda,\nu),$$

$$\lim_{\xi\to+\infty} e^{-\mu_2(\lambda,\nu)\xi}\mathbf{u}_2(\xi,\lambda,\nu) = \zeta_2(\lambda,\nu),$$

$$\lim_{\xi\to+\infty} e^{-\mu_3(\lambda,\nu)\xi}\mathbf{u}_3(\xi,\lambda,\nu) = \zeta_3(\lambda,\nu),$$

$$\lim_{\xi\to+\infty} e^{-\mu_4(\lambda,\nu)\xi}\mathbf{u}_4(\xi,\lambda,\nu) = \zeta_4(\lambda,\nu),$$

$$\lim_{\xi\to-\infty} e^{-\mu_5(\lambda,\nu)\xi}\mathbf{u}_5(\xi,\lambda,\nu) = \zeta_5(\lambda,\nu),$$

$$\lim_{\xi\to-\infty} e^{-\mu_6(\lambda,\nu)\xi}\mathbf{u}_6(\xi,\lambda,\nu) = \zeta_6(\lambda,\nu),$$

$$\lim_{\xi\to-\infty} e^{-\mu_7(\lambda,\nu)\xi}\mathbf{u}_7(\xi,\lambda,\nu) = \zeta_7(\lambda,\nu),$$

$$\lim_{\xi\to-\infty} e^{-\mu_8(\lambda,\nu)\xi}\mathbf{u}_8(\xi,\lambda,\nu) = \zeta_8(\lambda,\nu)\,. \tag{6.13}$$

The existence of the functions $\mathbf{u}_j$ follows from asymptotic theory that is well documented in the book of Coppel [52]. In the Evans function literature they are now sometimes called "Jost solutions", due to their similarity to constructions in quantum mechanics (e.g. Chapter 9 of Kapitula and Promislow [86], and Gesztesy et al. [69]).

The Evans function is defined in terms of the Jost functions to be

$$D(\lambda,\nu)\varpi = e^{-(\lambda\tau_1+\mathrm{i}\nu\tau_2)\xi}\mathbf{u}_1(\xi,\lambda,\nu)\wedge\ldots\wedge\mathbf{u}_8(\xi,\lambda,\nu)\,,\quad (\lambda,\nu)\in\Lambda\,, \tag{6.14}$$

with τ_1, τ_2 defined in (6.10). The function $D(\lambda,\nu)$ is independent of ξ and, with the hypothesis (**H6**), it is an analytic function of λ for all $(\lambda,\nu)\in\Lambda$ [4, 86].

The formal adjoint of (6.3) is

$$W_\xi = -\mathbf{A}(\xi,\lambda,\nu)^H W\,,\quad W\in V^*\,,\quad (\lambda,\nu)\in\Lambda\,, \tag{6.15}$$

where the superscript H denotes complex conjugate transpose. Solutions of that adjoint equation, paired with the original equation, have the property that

$$\frac{\mathrm{d}}{\mathrm{d}\xi}\overline{W}\cdot\mathbf{u} = 0\,,\quad \forall(\lambda,\nu)\in\Lambda\,. \tag{6.16}$$

A re-formulation of the adjoint equation, that takes into consideration the symplectic structure, is obtained by pre-multiplying (6.15) by $\mathbf{J}(c,\ell)^{-1}$,

$$\begin{aligned}(\mathbf{J}(c,\ell)^{-1}W)_\xi &= -\mathbf{J}(c,\ell)^{-1}\mathbf{A}(\xi,\lambda,\nu)^H W\\ &= \mathbf{J}(c,\ell)^{-1}[\mathbf{B}(\xi,c,\ell)+\overline{\lambda}\mathbf{M}-\mathrm{i}\nu\mathbf{N}](\mathbf{J}(c,\ell)^{-1}W)\,.\end{aligned}$$

Conjugating this equation gives

$$(\mathbf{J}(c,\ell)^{-1}\overline{W})_\xi = \mathbf{J}(c,\ell)^{-1}[\mathbf{B}(\xi,c,\ell)+\lambda\mathbf{M}+\mathrm{i}\nu\mathbf{N}](\mathbf{J}(c,\ell)^{-1}\overline{W})$$

or

$$(\mathbf{J}(c,\ell)^{-1}\overline{W})_\xi = \mathbf{A}(\xi,-\lambda,-\nu)(\mathbf{J}(c,\ell)^{-1}\overline{W})\,.$$

A natural definition of the solutions of the adjoint equation is then

$$\mathbf{w}(\xi,\lambda,\nu) = \mathbf{J}(c,\ell)^{-1}\overline{W(\xi,\lambda,\nu)}\,.$$

Viewing $\mathbf{J} : V \to V^*$, we can view $\mathbf{w} \in V$. The functions $\mathbf{w}(\xi,\lambda,\nu)$ are then analytic and satisfy

$$\mathbf{w}_\xi = \mathbf{A}(\xi,-\lambda,-\nu)\mathbf{w}\,,\quad \mathbf{w}\in V\,,\quad (\lambda,\nu)\in\Lambda\,. \tag{6.17}$$

Solutions of this equation paired with solutions of (6.3) are independent of ξ (6.16). Hence the property (6.16) can now be expressed in terms of the symplectic form,

$$\frac{\mathrm{d}}{\mathrm{d}\xi}\mathbf{\Omega}(\mathbf{w},\mathbf{u}) = 0\,. \tag{6.18}$$

There are eight solutions of the $\mathbf{w}$-equation (6.17) with the asymptotic properties

$$\lim_{\xi\to-\infty} e^{+\mu_1(\lambda,\nu)\xi}\mathbf{w}_1(\xi,\lambda,\nu) = \eta_1(\lambda,\nu),$$

$$\lim_{\xi\to-\infty} e^{+\mu_2(\lambda,\nu)\xi}\mathbf{w}_2(\xi,\lambda,\nu) = \eta_2(\lambda,\nu),$$

$$\lim_{\xi\to-\infty} e^{+\mu_3(\lambda,\nu)\xi}\mathbf{w}_3(\xi,\lambda,\nu) = \eta_3(\lambda,\nu),$$

$$\lim_{\xi\to-\infty} e^{+\mu_4(\lambda,\nu)\xi}\mathbf{w}_4(\xi,\lambda,\nu) = \eta_4(\lambda,\nu),$$

$$\lim_{\xi\to+\infty} e^{+\mu_5(\lambda,\nu)\xi}\mathbf{w}_5(\xi,\lambda,\nu) = \eta_5(\lambda,\nu),$$

$$\lim_{\xi\to+\infty} e^{+\mu_6(\lambda,\nu)\xi}\mathbf{w}_6(\xi,\lambda,\nu) = \eta_6(\lambda,\nu),$$

$$\lim_{\xi\to+\infty} e^{+\mu_7(\lambda,\nu)\xi}\mathbf{w}_7(\xi,\lambda,\nu)=\eta_7(\lambda,\nu),$$

$$\lim_{\xi\to+\infty} e^{+\mu_8(\lambda,\nu)\xi}\mathbf{w}_8(\xi,\lambda,\nu)=\eta_8(\lambda,\nu)\,. \tag{6.19}$$

Here η_j, $j=1,\ldots,8$ are eigenvectors associated with the adjoint system at infinity

$$[\mathbf{B}^\infty+\lambda\mathbf{M}+\mathrm{i}\nu\mathbf{N}+\mu_j\mathbf{J}(c,\ell)]\eta_j=0\,,\quad j=1,\ldots,8, \tag{6.20}$$

and they are normalised by

$$\mathbf{\Omega}(\eta_i,\zeta_j)=\delta_{ij}\,,\quad i,j=1,\ldots,8\,. \tag{6.21}$$

Before transforming the Evans function to a matrix definition, we introduce some of the necessary exterior algebra constructions. A basis for V is $\{\mathbf{e}_1,\ldots,\mathbf{e}_8\}$, which is not necessarily the standard basis, and a basis for V^* is $\{\mathbf{e}_1^*,\ldots,\mathbf{e}_8^*\}$ normalised via $\delta_{ij}=\langle\mathbf{e}_i^*,\mathbf{e}_j\rangle$, $i,j=1,\ldots,8$. Exterior algebra spaces $\bigwedge^k(V)$ and $\bigwedge^k(V^*)$, $k=1,2,\ldots,8$, are defined in the usual way with induced pairings, $[\![\cdot,\cdot]\!]_k$ given by

$$[\![\mathbf{a}_1\wedge\ldots\wedge\mathbf{a}_k,\mathbf{b}_1\wedge\ldots\wedge\mathbf{b}_k]\!]_k$$
$$=\det\begin{pmatrix}\langle\mathbf{a}_1,\mathbf{b}_1\rangle & \cdots & \langle\mathbf{a}_1,\mathbf{b}_k\rangle\\ \vdots & \ddots & \vdots\\ \langle\mathbf{a}_k,\mathbf{b}_1\rangle & \cdots & \langle\mathbf{a}_k,\mathbf{b}_k\rangle\end{pmatrix},\ \text{for}\,\mathbf{a}_j\in V^*,\ \mathbf{b}_j\in V,\ j=1,\ldots,k$$

for a decomposable element. Bilinearity is then used to extend this definition to arbitrary elements in $\bigwedge^k$. Note that the pairing $[\![\cdot,\cdot]\!]_1$ is equivalent to $\langle\cdot,\cdot\rangle$, viewed as a pairing between V and V^*. Although V is a complex vector space, the pairing between V^* and V does not involve conjugation (to be consistent with the definition of the dual vectors η_j in (6.20) where conjugation is built in). The volume form on V is (5.22) and on V^* it is

$$\varpi^*=\mathbf{e}_1^*\wedge\mathbf{e}_2^*\wedge\cdots\wedge\mathbf{e}_7^*\wedge\mathbf{e}_8^*\,.$$

The normalised basis then gives that $[\![\varpi^*,\varpi]\!]_8=1$. It will also be necessary to consider the equivalent bases

$$V=\mathrm{span}\{\zeta_1,\ldots,\zeta_8\}\quad\text{and}\quad V^*=\mathrm{span}\{\mathbf{J}(c,\ell)\eta_1,\ldots,\mathbf{J}(c,\ell)\eta_8\}\,, \tag{6.22}$$

where ζ_j and η_j are the eigenvectors defined in (6.11) and (6.20) respectively, with their associated eight-forms $\mathcal{V}(\lambda,\nu,c,\ell)$, defined by

$$\mathcal{V}(\lambda,\nu):=\zeta_1(\lambda,\nu)\wedge\cdots\wedge\zeta_8(\lambda,\nu)=K(\lambda,\nu)\varpi\,, \tag{6.23}$$

with the dependence on c, ℓ suppressed for brevity, for some $K(\lambda, \nu) \neq 0$, for all $c, \ell \in \mathscr{C}$ and $\lambda, \nu \in \Lambda$. The volume form dual to $\mathcal{V}$ is

$$\mathcal{V}^*(\lambda, \nu) := \mathbf{J}(c, \ell)\eta_1(\lambda, \nu) \wedge \cdots \wedge \mathbf{J}(c, \ell)\eta_8(\lambda, \nu) . \tag{6.24}$$

Due to (6.21), the volume forms satisfy $[\![\mathcal{V}^*, \mathcal{V}]\!]_8 = 1$. With respect to the pairings $[\![\cdot, \cdot]\!]_k$ the interior product (here denoted by $\lrcorner$) is the adjoint operator associated with the wedge product so that

$$[\![\mathbf{a} \lrcorner\, \mathbf{b}, \mathbf{c}]\!]_{k-1} = [\![\mathbf{b}, \mathbf{a} \wedge \mathbf{c}]\!]_k , \quad \forall\, \mathbf{a} \in V , \ \mathbf{b} \in \textstyle\bigwedge^k(V^*) , \ \mathbf{c} \in \textstyle\bigwedge^{k-1}(V) . \tag{6.25}$$

It will be necessary to use the following recursive definition:

$$(\mathbf{a} \wedge \mathbf{b}) \lrcorner\, \mathbf{c} := \mathbf{b} \lrcorner\, (\mathbf{a} \lrcorner\, \mathbf{c}) , \quad \forall\, \mathbf{a}, \mathbf{b} \in V , \ \mathbf{c} \in \textstyle\bigwedge^k(V^*) . \tag{6.26}$$

The following identity then follows from (6.25) and (6.26):

$$[\![\mathbf{a} \lrcorner\, \mathbf{b}, \mathbf{c}]\!]_4 = [\![\mathbf{b}, \mathbf{a} \wedge \mathbf{c}]\!]_8 , \quad \forall\, \mathbf{a}, \mathbf{c} \in \textstyle\bigwedge^4(V) , \ \mathbf{b} \in \textstyle\bigwedge^8(V^*) . \tag{6.27}$$

The symplectic structure does not appear explicitly in the formula for the Evans function (6.14), but will appear explicitly in the matrix-definition. The following proposition shows how symplecticity enters into the transformation.

Proposition 6.1 *The four-form $\zeta_1 \wedge \zeta_2 \wedge \zeta_3 \wedge \zeta_4 \in \bigwedge^4(V)$ transforms to a four-form in $\bigwedge^4(V^*)$ with symplectic structure via*

$$(\zeta_1 \wedge \zeta_2 \wedge \zeta_3 \wedge \zeta_4) \lrcorner\, \mathcal{V}^* = \mathbf{J}(c, \ell)\eta_5 \wedge \mathbf{J}(c, \ell)\eta_6 \wedge \mathbf{J}(c, \ell)\eta_7 \wedge \mathbf{J}(c, \ell)\eta_8 . \tag{6.28}$$

Proof The basis $\{\mathbf{J}\eta_1, \ldots, \mathbf{J}\eta_8\}$ of V^* extends to a basis for $\bigwedge^4(V^*)$ with basis elements

$$\mathbf{J}\eta_i \wedge \mathbf{J}\eta_j \wedge \mathbf{J}\eta_k \wedge \mathbf{J}\eta_l ,$$

for $i, j, k, l = 1, \ldots, 8$ and with $i < j < k < l$. Expand the left-hand side of (6.28) in terms of this basis

$$\sum_{i,j,k,l} \mathcal{P}_{i,j,k,l} \mathbf{J}\eta_i \wedge \mathbf{J}\eta_j \wedge \mathbf{J}\eta_k \wedge \mathbf{J}\eta_l = (\zeta_1 \wedge \zeta_2 \wedge \zeta_3 \wedge \zeta_4) \lrcorner\, \mathcal{V}^* . \tag{6.29}$$

Take an arbitrary basis element in $\bigwedge^4(V)$

$$\zeta_r \wedge \zeta_s \wedge \zeta_t \wedge \zeta_u \in \textstyle\bigwedge^4(V) ,$$

and pair with both sides of (6.29):

$$\sum_{i,j,k,l} \mathcal{P}_{i,j,k,l} [\![\mathbf{J}\eta_i \wedge \mathbf{J}\eta_j \wedge \mathbf{J}\eta_k \wedge \mathbf{J}\eta_l , \zeta_r \wedge \zeta_s \wedge \zeta_t \wedge \zeta_u]\!]_4$$

$$= [\![(\zeta_1 \wedge \zeta_2 \wedge \zeta_3 \wedge \zeta_4) \lrcorner\, \mathcal{V}^*, \zeta_r \wedge \zeta_s \wedge \zeta_t \wedge \zeta_u]\!]_4 \,.$$

The right-hand side transforms to

$$[\![(\zeta_1 \wedge \zeta_2 \wedge \zeta_3 \wedge \zeta_4) \lrcorner\, \mathcal{V}^*, \zeta_r \wedge \zeta_s \wedge \zeta_t \wedge \zeta_u]\!]_4 = [\![\mathcal{V}^*, \zeta_1 \wedge \zeta_2 \wedge \zeta_3 \wedge \zeta_4 \wedge \zeta_r \wedge \zeta_s \wedge \zeta_t \wedge \zeta_u]\!]_8 \,.$$

The right-hand side is zero except when $r, s, t, u = 5, 6, 7, 8$ in which case the right-hand side is unity. Hence

$$\sum_{i,j,k,l} \mathcal{P}_{i,j,k,l} [\![\mathbf{J}\eta_i \wedge \mathbf{J}\eta_j \wedge \mathbf{J}\eta_k \wedge \mathbf{J}\eta_l , \zeta_5 \wedge \zeta_6 \wedge \zeta_7 \wedge \zeta_8]\!]_4 = 1 \,.$$

The left-hand side is zero unless $i, j, k, l = 5, 6, 7, 8$, giving zero for all $\mathcal{P}_{i,j,k,l}$ except for $\mathcal{P}_{5,6,7,8}$ which is unity, completing the proof of (6.28). ■

6.2 Matrix Definition of the Evans Function

We are now in a position to state a proof of the main result relating the two forms of the Evans function.

Theorem 6.2 *With the orientation* (5.22) *and the normalisations* (6.21), *the Evans function* (6.14) *can be transformed to the representation*

$$D(\lambda, \nu) = \frac{1}{2} \det \begin{bmatrix} \mathbf{\Omega}(\mathbf{w}_5, \mathbf{u}_5) & \mathbf{\Omega}(\mathbf{w}_5, \mathbf{u}_6) & \mathbf{\Omega}(\mathbf{w}_5, \mathbf{u}_7) & \mathbf{\Omega}(\mathbf{w}_5, \mathbf{u}_8) \\ \mathbf{\Omega}(\mathbf{w}_6, \mathbf{u}_5) & \mathbf{\Omega}(\mathbf{w}_6, \mathbf{u}_6) & \mathbf{\Omega}(\mathbf{w}_6, \mathbf{u}_7) & \mathbf{\Omega}(\mathbf{w}_6, \mathbf{u}_8) \\ \mathbf{\Omega}(\mathbf{w}_7, \mathbf{u}_5) & \mathbf{\Omega}(\mathbf{w}_7, \mathbf{u}_6) & \mathbf{\Omega}(\mathbf{w}_7, \mathbf{u}_7) & \mathbf{\Omega}(\mathbf{w}_7, \mathbf{u}_8) \\ \mathbf{\Omega}(\mathbf{w}_8, \mathbf{u}_5) & \mathbf{\Omega}(\mathbf{w}_8, \mathbf{u}_6) & \mathbf{\Omega}(\mathbf{w}_8, \mathbf{u}_7) & \mathbf{\Omega}(\mathbf{w}_8, \mathbf{u}_8) \end{bmatrix} . \tag{6.30}$$

Proof The difficult parts are firstly bringing in the symplectic structure, and the key result for this is (6.28), and secondly synchronising the asymptotics of the two definitions. The strategy is similar to the construction in four dimensions, without a metric, in Chapter 4 of Burchell [34], and Appendix A of Burchell and Bridges [35]. Hence we will just give a sketch here.

The starting point of the proof is the Evans function in the form (6.14), relative to the orientation ϖ which we will need to be shifted to the orientation $\mathcal{V}$. The interior product is used to transform it as follows, suppressing the arguments to simplify notation,

$$
\begin{aligned}
D(\lambda,\nu)\,\varpi &= e^{-(\lambda\tau_1+\mathrm{i}\nu\tau_2)\xi}\mathbf{u}_1\wedge\cdots\wedge\mathbf{u}_8\\
&= [\![\varpi^*, e^{-(\lambda\tau_1+\mathrm{i}\nu\tau_2)\xi}\mathbf{u}_1\wedge\cdots\wedge\mathbf{u}_8]\!]_8\varpi\\
&= [\![\mathcal{V}^*(\lambda,\nu), e^{-\tau(c)\lambda\xi}\mathbf{u}_1\wedge\cdots\wedge\mathbf{u}_8]\!]_8\mathcal{V}(\lambda,\nu)\\
&= [\![e^{-(\lambda\tau_1+\mathrm{i}\nu\tau_2)\xi}(\mathbf{u}_1\wedge\cdots\wedge\mathbf{u}_4)\,\lrcorner\,\mathcal{V}^*(\lambda,\nu), \mathbf{u}_5\wedge\cdots\wedge\mathbf{u}_8]\!]_4\mathcal{V}(\lambda,\nu)\,,
\end{aligned}
\tag{6.31}
$$

using $1=[\![\varpi^*,\varpi]\!]_8=[\![\mathcal{V}^*,\mathcal{V}]\!]_8$ and (6.27). Define the left argument in the pairing as

$$
\mathbf{W}(\xi,\lambda,\nu)=e^{-(\lambda\tau_1+\mathrm{i}\nu\tau_2)\xi}\big(\mathbf{u}_1(\xi,\lambda,\nu)\wedge\cdots\wedge\mathbf{u}_4(\xi,\lambda,\nu)\big)\,\lrcorner\,\mathcal{V}^*\in{\textstyle\bigwedge}^4(V^*)\,. \tag{6.32}
$$

Differentiate it with respect to ξ

$$
\mathbf{W}_\xi=-\left(\lambda\tau_1+\mathrm{i}\nu\tau_2\right)\mathbf{W}+e^{-(\lambda\tau_1+\mathrm{i}\nu\tau_2)\xi}\left(\mathbf{A}^{(4)}(\mathbf{u}_1\wedge\cdots\wedge\mathbf{u}_4)\right)\,\lrcorner\,\mathcal{V}^*\,, \tag{6.33}
$$

using the definition

$$
\begin{aligned}
&\mathbf{A}^{(4)}(\mathbf{a}_1\wedge\cdots\wedge\mathbf{a}_4)\\
&\quad:=\mathbf{A}\mathbf{a}_1\wedge\cdots\wedge\mathbf{a}_4+\cdots+\mathbf{a}_1\wedge\cdots\wedge\mathbf{A}\mathbf{a}_4\,,\qquad \text{for any } \mathbf{a}_1,\ldots,\mathbf{a}_4\in V\,,
\end{aligned}
\tag{6.34}
$$

and the governing equation for $\mathbf{u}_j$ in (6.12). Now look at the second term in (6.33). It can be split using the identity

$$
\begin{aligned}
&\mathbf{A}^{(4)}\big(\mathbf{a}_1\wedge\cdots\wedge\mathbf{a}_4\big)\wedge\mathbf{b}_1\wedge\cdots\wedge\mathbf{b}_4+\mathbf{a}_1\wedge\cdots\wedge\mathbf{a}_4\wedge\mathbf{A}^{(4)}\big(\mathbf{b}_1\wedge\cdots\wedge\mathbf{b}_4\big)\\
&\quad=\mathrm{Tr}(\mathbf{A})\mathbf{a}_1\wedge\cdots\wedge\mathbf{a}_4\wedge\mathbf{b}_1\wedge\cdots\wedge\mathbf{b}_4\,.
\end{aligned}
$$

Let $\mathbf{a}_1,\ldots,\mathbf{a}_4\in V$ be arbitrary and pair $\mathbf{a}_1\wedge\cdots\wedge\mathbf{a}_4\in\bigwedge^4(V)$ with the operator in the second term on the right-hand side of (6.33),

$$
\begin{aligned}
&[\![\left(\mathbf{A}^{(4)}(\mathbf{u}_1\wedge\cdots\wedge\mathbf{u}_4)\right)\,\lrcorner\,\mathcal{V}^*,\mathbf{a}_1\wedge\cdots\wedge\mathbf{a}_4]\!]_4\\
&\quad=[\![\mathcal{V}^*,\mathbf{A}^{(4)}\big(\mathbf{u}_1\wedge\cdots\wedge\mathbf{u}_4\big)\wedge\mathbf{a}_1\wedge\cdots\wedge\mathbf{a}_4]\!]_8\\
&\quad=\mathrm{Tr}(\mathbf{A})[\![\mathcal{V}^*,\mathbf{u}_1\wedge\cdots\wedge\mathbf{u}_4\wedge\mathbf{a}_1\wedge\cdots\wedge\mathbf{a}_4]\!]_8\\
&\qquad-[\![\mathcal{V}^*,\mathbf{u}_1\wedge\cdots\wedge\mathbf{u}_4\wedge\mathbf{A}^{(4)}\big(\mathbf{a}_1\wedge\cdots\wedge\mathbf{a}_4\big)]\!]_8\\
&\quad=\mathrm{Tr}(\mathbf{A})[\![\big(\mathbf{u}_1\wedge\cdots\wedge\mathbf{u}_4\big)\,\lrcorner\,\mathcal{V}^*,\mathbf{a}_1\wedge\cdots\wedge\mathbf{a}_4]\!]_4\\
&\qquad-[\![\big(\mathbf{u}_1\wedge\cdots\wedge\mathbf{u}_4\big)\,\lrcorner\,\mathcal{V}^*,\mathbf{A}^{(4)}\big(\mathbf{a}_1\wedge\cdots\mathbf{a}_4\big)]\!]_4\\
&\quad=(\lambda\tau_1+\mathrm{i}\nu\tau_2)[\![\big(\mathbf{u}_1\wedge\cdots\wedge\mathbf{u}_4\big)\,\lrcorner\,\mathcal{V}^*,\mathbf{a}_1\wedge\cdots\mathbf{a}_4]\!]_4\\
&\qquad-[\![(\mathbf{A}^{(4)})^T\left[\big(\mathbf{u}_1\wedge\cdots\wedge\mathbf{u}_4\big)\,\lrcorner\,\mathcal{V}^*\right],\mathbf{a}_1\wedge\cdots\wedge\mathbf{a}_4]\!]_4\,,
\end{aligned}
$$

using (6.27) and (6.34). Since this holds for any $\mathbf{a}_1, \ldots, \mathbf{a}_4 \in V$ we get that

$$\begin{aligned}
&\mathbf{A}^{(4)}(\mathbf{u}_1 \wedge \cdots \wedge \mathbf{u}_4) \rfloor \mathcal{V}^* \\
&\quad = (\lambda\tau_1 + \mathrm{i}\nu\tau_2)(\mathbf{u}_1 \wedge \cdots \wedge \mathbf{u}_4) \rfloor \mathcal{V}^* - (\mathbf{A}^{(4)})^T \left[(\mathbf{u}_1 \wedge \cdots \wedge \mathbf{u}_4) \rfloor \mathcal{V}^*\right].
\end{aligned}$$

Substitute into (6.33):

$$\begin{aligned}
\mathbf{W}_\xi &= -(\lambda\tau_1 + \mathrm{i}\nu\tau_2)\mathbf{W} + e^{-(\lambda\tau_1+\mathrm{i}\nu\tau_2)\xi}\left(\mathbf{A}^{(4)}(\mathbf{u}_1 \wedge \cdots \wedge \mathbf{u}_4)\right) \rfloor \mathcal{V}^* \\
&= -(\lambda\tau_1 + \mathrm{i}\nu\tau_2)\mathbf{W} + \left((\lambda\tau_1 + \mathrm{i}\nu\tau_2)\mathbf{W} - (\mathbf{A}^{(4)})^T\mathbf{W}\right),
\end{aligned}$$

or

$$\mathbf{W}_\xi = -\left(\mathbf{A}^{(4)}(\xi, \lambda, \nu)\right)^T \mathbf{W}, \quad \mathbf{W} \in \bigwedge\nolimits^4(V^*). \tag{6.35}$$

This equation shows that $\mathbf{W}$ satisfies the adjoint equation on $\bigwedge^4$.

Now compare this equation for $\mathbf{W}$ with the adjoint equation on $\bigwedge^4$ derived directly; let

$$\Upsilon(\xi, \lambda, \nu) = \mathbf{J}(c, \ell)\mathbf{w}_5(\xi, \lambda, \nu) \wedge \cdots \wedge \mathbf{J}(c, \ell)\mathbf{w}_8(\xi, \lambda, \nu), \tag{6.36}$$

and differentiate with respect to ξ, dropping the argument on $\mathbf{J}$,

$$\begin{aligned}
\Upsilon_\xi &= \mathbf{J}(\mathbf{w}_5)_\xi \wedge \cdots \wedge \mathbf{J}\mathbf{w}_8 + \cdots + \mathbf{J}\mathbf{w}_5 \wedge \cdots \wedge \mathbf{J}(\mathbf{w}_8)_\xi \\
&= \mathbf{J}\mathbf{A}(\xi, -\lambda - \nu)\mathbf{w}_5 \wedge \cdots \wedge \mathbf{J}\mathbf{w}_8 + \cdots + \mathbf{J}\mathbf{w}_5 \wedge \cdots \wedge \mathbf{J}\mathbf{A}(\xi, -\lambda, -\nu)\mathbf{w}_8 \\
&= (\mathbf{B}(\xi, c, \ell) + \lambda\mathbf{M} + \mathrm{i}\nu\mathbf{N})\mathbf{w}_5 \wedge \cdots \wedge \mathbf{J}\mathbf{w}_8 + \cdots + \mathbf{J}\mathbf{w}_5 \wedge \cdots \\
&\qquad \wedge(\mathbf{B}(\xi, c, \ell) + \lambda\mathbf{M} + \mathrm{i}\nu\mathbf{N})\mathbf{w}_8 \\
&= -\mathbf{A}(\xi, \lambda, \nu)^T\mathbf{J}\mathbf{w}_5 \wedge \cdots \wedge \mathbf{J}\mathbf{w}_8 - \cdots - \mathbf{J}\mathbf{w}_5 \wedge \cdots \wedge \mathbf{A}(\xi, \lambda, \nu)^T\mathbf{J}\mathbf{w}_8,
\end{aligned}$$

using the governing equation for $\mathbf{w}$ in (6.17), which simplifies to

$$\Upsilon_\xi = -\left(\mathbf{A}^{(4)}(\xi, \lambda, \nu)\right)^T \Upsilon, \quad \Upsilon \in \bigwedge\nolimits^4(V^*). \tag{6.37}$$

It is clear from (6.35) and (6.37) that $\mathbf{W}$ and Υ satisfy the same adjoint equation on $\bigwedge^4(V^*)$. However, to show equality; that is,

$$e^{-(\lambda\tau_1+\mathrm{i}\nu\tau_2)\xi}(\mathbf{u}_1 \wedge \cdots \wedge \mathbf{u}_4) \rfloor \mathcal{V}^* = \mathbf{J}\mathbf{w}_5 \wedge \cdots \wedge \mathbf{J}\mathbf{w}_8, \tag{6.38}$$

it is necessary to study their asymptotics. Using the asymptotic properties of $\mathbf{w}_5, \ldots, \mathbf{w}_8$ in (6.19),

$$\lim_{\xi\to+\infty} e^{(\mu_5+\cdots+\mu_8)\xi}\Upsilon(\xi,\lambda) = \lim_{\xi\to+\infty} \left(e^{\mu_5\xi}\mathbf{J}\mathbf{w}_5\right)\wedge\cdots\wedge\left(e^{\mu_8\xi}\mathbf{J}\mathbf{w}_8\right)$$
$$= \mathbf{J}\eta_5\wedge\cdots\wedge\mathbf{J}\eta_8\,. \tag{6.39}$$

On the other hand, using $(\lambda\tau_1 + \mathtt{i}\nu\tau_2) = \mu_1 + \cdots + \mu_8$,

$$e^{(\mu_5+\cdots+\mu_8)\xi}\mathbf{W}(\xi,\lambda,\nu) = e^{-(\mu_1+\cdots+\mu_4)\xi}(\mathbf{u}_1\wedge\cdots\wedge\mathbf{u}_4)\,\lrcorner\,\mathcal{V}^*\,.$$

Taking the limit $\xi\to+\infty$ and using the asymptotics of $\mathbf{u}_1,\ldots,\mathbf{u}_4$,

$$\lim_{\xi\to+\infty} e^{(\mu_5+\cdots+\mu_8)\xi}\mathbf{W}(\xi,\lambda,\nu) = \lim_{\xi\to+\infty} e^{-(\mu_1+\cdots+\mu_4)\xi}(\mathbf{u}_1\wedge\cdots\wedge\mathbf{u}_4)\,\lrcorner\,\mathcal{V}^*$$
$$= (\zeta_1\wedge\cdots\wedge\zeta_4)\,\lrcorner\,\mathcal{V}^*\,. \tag{6.40}$$

Applying formula (6.28) shows that (6.39) and (6.40) have the same asymptotics and are therefore equal: $\mathbf{W}(\xi,\lambda,\nu) = \Upsilon(\xi,\lambda,\nu)$.

Now apply the identity $\mathbf{W} = \Upsilon$ to the Evans function in (6.31),

$$D(\lambda,\nu)\,\varpi = [\![e^{-(\lambda\tau_1+\mathtt{i}\nu\tau_2)\xi}(\mathbf{u}_1\wedge\cdots\wedge\wedge\mathbf{u}_4)\,\lrcorner\,\mathcal{V}^*, \mathbf{u}_5\wedge\cdots\wedge\mathbf{u}_8]\!]_4\mathcal{V}$$
$$= [\![\mathbf{J}(c,\ell)\mathbf{w}_5\wedge\cdots\wedge\mathbf{J}(c,\ell)\mathbf{w}_8, \mathbf{u}_5\wedge\cdots\wedge\mathbf{u}_8]\!]_4\,\mathcal{V}$$
$$= \det\begin{bmatrix} \mathbf{\Omega}(\mathbf{w}_5,\mathbf{u}_5) & \mathbf{\Omega}(\mathbf{w}_5,\mathbf{u}_6) & \mathbf{\Omega}(\mathbf{w}_5,\mathbf{u}_7) & \mathbf{\Omega}(\mathbf{w}_5,\mathbf{u}_8) \\ \mathbf{\Omega}(\mathbf{w}_6,\mathbf{u}_5) & \mathbf{\Omega}(\mathbf{w}_6,\mathbf{u}_6) & \mathbf{\Omega}(\mathbf{w}_7,\mathbf{u}_7) & \mathbf{\Omega}(\mathbf{w}_8,\mathbf{u}_8) \\ \mathbf{\Omega}(\mathbf{w}_7,\mathbf{u}_5) & \mathbf{\Omega}(\mathbf{w}_7,\mathbf{u}_6) & \mathbf{\Omega}(\mathbf{w}_7,\mathbf{u}_7) & \mathbf{\Omega}(\mathbf{w}_7,\mathbf{u}_8) \\ \mathbf{\Omega}(\mathbf{w}_8,\mathbf{u}_5) & \mathbf{\Omega}(\mathbf{w}_8,\mathbf{u}_6) & \mathbf{\Omega}(\mathbf{w}_8,\mathbf{u}_7) & \mathbf{\Omega}(\mathbf{w}_8,\mathbf{u}_8) \end{bmatrix}\mathcal{V}\,, \tag{6.41}$$

$$= K(\lambda,\nu)\det\begin{bmatrix} \mathbf{\Omega}(\mathbf{w}_5,\mathbf{u}_5) & \mathbf{\Omega}(\mathbf{w}_5,\mathbf{u}_6) & \mathbf{\Omega}(\mathbf{w}_5,\mathbf{u}_7) & \mathbf{\Omega}(\mathbf{w}_5,\mathbf{u}_8) \\ \mathbf{\Omega}(\mathbf{w}_6,\mathbf{u}_5) & \mathbf{\Omega}(\mathbf{w}_6,\mathbf{u}_6) & \mathbf{\Omega}(\mathbf{w}_7,\mathbf{u}_7) & \mathbf{\Omega}(\mathbf{w}_8,\mathbf{u}_8) \\ \mathbf{\Omega}(\mathbf{w}_7,\mathbf{u}_5) & \mathbf{\Omega}(\mathbf{w}_7,\mathbf{u}_6) & \mathbf{\Omega}(\mathbf{w}_7,\mathbf{u}_7) & \mathbf{\Omega}(\mathbf{w}_7,\mathbf{u}_8) \\ \mathbf{\Omega}(\mathbf{w}_8,\mathbf{u}_5) & \mathbf{\Omega}(\mathbf{w}_8,\mathbf{u}_6) & \mathbf{\Omega}(\mathbf{w}_8,\mathbf{u}_7) & \mathbf{\Omega}(\mathbf{w}_8,\mathbf{u}_8) \end{bmatrix}\varpi\,, \tag{6.42}$$

using $\mathcal{V}(\lambda,\nu) = K(\lambda,\nu)\varpi$. ∎

Remark 5 The factor of $\frac{1}{2}$ will arise below, based on the concept of equivalent Evans functions. Two representations $D^A(\lambda)$ and $D^B(\lambda)$ of an Evans function are equivalent if there exists a non-vanishing analytic function $C(\lambda)$ such that $D^A(\lambda) = C(\lambda)D^B(\lambda)$ for all $\lambda\in\Lambda$. When orientation of $D^A(\lambda)$ and $D^B(\lambda)$ along the real axis is of interest, then $C(\lambda)\big|_{\lambda\in\mathbb{R}}$ is required to be real.

Remark 6 Noting that $\mathcal{V}(\lambda, \nu) = K(\lambda, \nu)\varpi$ in (6.30), the explicit formula for the scalar-valued function $D(\lambda, \nu)$ is

$$D(\lambda, \nu) = K(\lambda, \nu) \det \begin{bmatrix} \boldsymbol{\Omega}(\mathbf{w}_5, \mathbf{u}_5) & \boldsymbol{\Omega}(\mathbf{w}_5, \mathbf{u}_6) & \boldsymbol{\Omega}(\mathbf{w}_5, \mathbf{u}_7) & \boldsymbol{\Omega}(\mathbf{w}_5, \mathbf{u}_8) \\ \boldsymbol{\Omega}(\mathbf{w}_6, \mathbf{u}_5) & \boldsymbol{\Omega}(\mathbf{w}_6, \mathbf{u}_6) & \boldsymbol{\Omega}(\mathbf{w}_7, \mathbf{u}_7) & \boldsymbol{\Omega}(\mathbf{w}_8, \mathbf{u}_8) \\ \boldsymbol{\Omega}(\mathbf{w}_7, \mathbf{u}_5) & \boldsymbol{\Omega}(\mathbf{w}_7, \mathbf{u}_6) & \boldsymbol{\Omega}(\mathbf{w}_7, \mathbf{u}_7) & \boldsymbol{\Omega}(\mathbf{w}_7, \mathbf{u}_8) \\ \boldsymbol{\Omega}(\mathbf{w}_8, \mathbf{u}_5) & \boldsymbol{\Omega}(\mathbf{w}_8, \mathbf{u}_6) & \boldsymbol{\Omega}(\mathbf{w}_8, \mathbf{u}_7) & \boldsymbol{\Omega}(\mathbf{w}_8, \mathbf{u}_8) \end{bmatrix}.$$

However, as noted in (6.23), $K(\lambda, \nu)$ is non-zero for all $(c, \ell) \in \mathscr{C}$ and all $(\lambda, \nu) \in \Lambda$. Hence by taking the transforming function to be

$$C(\lambda) = \frac{1}{2K(\lambda, \nu)},$$

the equivalent Evans function

$$D(\lambda, \nu) := \frac{1}{2} \det \begin{bmatrix} \boldsymbol{\Omega}(\mathbf{w}_5, \mathbf{u}_5) & \boldsymbol{\Omega}(\mathbf{w}_5, \mathbf{u}_6) & \boldsymbol{\Omega}(\mathbf{w}_5, \mathbf{u}_7) & \boldsymbol{\Omega}(\mathbf{w}_5, \mathbf{u}_8) \\ \boldsymbol{\Omega}(\mathbf{w}_6, \mathbf{u}_5) & \boldsymbol{\Omega}(\mathbf{w}_6, \mathbf{u}_6) & \boldsymbol{\Omega}(\mathbf{w}_7, \mathbf{u}_7) & \boldsymbol{\Omega}(\mathbf{w}_8, \mathbf{u}_8) \\ \boldsymbol{\Omega}(\mathbf{w}_7, \mathbf{u}_5) & \boldsymbol{\Omega}(\mathbf{w}_7, \mathbf{u}_6) & \boldsymbol{\Omega}(\mathbf{w}_7, \mathbf{u}_7) & \boldsymbol{\Omega}(\mathbf{w}_7, \mathbf{u}_8) \\ \boldsymbol{\Omega}(\mathbf{w}_8, \mathbf{u}_5) & \boldsymbol{\Omega}(\mathbf{w}_8, \mathbf{u}_6) & \boldsymbol{\Omega}(\mathbf{w}_8, \mathbf{u}_7) & \boldsymbol{\Omega}(\mathbf{w}_8, \mathbf{u}_8) \end{bmatrix} \tag{6.43}$$

is obtained. Here and henceforth, it is this function that will be used as the "Evans function". The extra factor $\frac{1}{2}$ simplifies some of the derivative calculations in Chap. 7.

Remark 7 One consequence of equivalence is that $D(\lambda, \nu)$ and $-D(\lambda, \nu)$ are equivalent Evans functions. There is no natural sign of the Evans function, and so orientations and fixing the signs of quantities are important. On the other hand, the relative sign

$$\text{sign}\Big(D(\lambda_1, \nu)D(\lambda_2, \nu)\Big), \quad \lambda_2 > \lambda_1,$$

with $\lambda_1, \lambda_2 \in \mathbb{R}$, along with the intermediate value theorem, can be used to capture unstable eigenvalues. We will also introduce another instability condition that is independent of the orientation of $D(\lambda, \nu)$ which is based on small ν (see Corollary 7.2 in Chap. 7).

Remark 8 The above definition of the Evans function is invariant under and permutation of $\{5, 6, 7, 8\}$. This follows from the permutation property of matrix columns and rows. This property will be useful in the proof of the derivative formula as it shows that it does not matter whether $\widehat{Z}_\xi$ is asymptotic to the direction ζ_5, ζ_6, ζ_7, or ζ_8 as $\xi \to -\infty$.

6.3 Small (λ, ν) Asymptotics and Derivatives

We record some of the limits of the $\mathbf{u}_j$ and $\mathbf{w}_j$, in preparation for derivatives of the Evans function,

$$\begin{aligned}
&\lim_{\lambda,\nu\to 0} \mathbf{u}_5(\xi,\lambda,\nu) = \frac{1}{\chi^-}\widehat{Z}_\xi\,, \qquad && \lim_{\lambda,\nu\to 0} \mathbf{u}_6(\xi,\lambda,\nu) = \mathbf{a}^-(\xi)\,,\\
&\lim_{\lambda,\nu\to 0} \mathbf{u}_7(\xi,\lambda,\nu) = \mathbf{b}^-(\xi)\,, && \lim_{\lambda,\nu\to 0} \mathbf{u}_8(\xi,\lambda,\nu) = \mathbf{c}^-(\xi)\,.
\end{aligned} \tag{6.44}$$

The first limit is due to the fact that

$$e^{-\mu_5(\lambda,\nu)\xi}\mathbf{u}_5 \to \zeta_5 \quad \text{but} \quad e^{-\mu_5(\lambda,\nu)\xi}\widehat{Z}_\xi \to \chi^-\zeta_5\,,$$

using (6.13) and (5.29). The λ, ν-limit of the other terms in (6.44) follow from (6.13) and (5.1). Similarly, the adjoint solutions have $\lambda, \nu \to 0$ limits,

$$\begin{aligned}
&\lim_{\lambda,\nu\to 0} \mathbf{w}_5(\xi,\lambda,\nu) = \frac{1}{\chi^+}\widehat{Z}_\xi\,, \qquad && \lim_{\lambda,\nu\to 0} \mathbf{w}_6(\xi,\lambda,\nu) = \mathbf{a}^+(\xi)\,,\\
&\lim_{\lambda,\nu\to 0} \mathbf{w}_7(\xi,\lambda,\nu) = \mathbf{b}^+(\xi)\,, && \lim_{\lambda,\nu\to 0} \mathbf{w}_8(\xi,\lambda,\nu) = \mathbf{c}^+(\xi)\,.
\end{aligned} \tag{6.45}$$

The form of the solutions $\mathbf{u}_j$ and adjoint solutions $\mathbf{w}_j$ and their limits (6.13), (6.19), (6.44), (6.45), and the form of the Evans function (6.30) together constitute hypothesis (**H7**).

The last thing that is required before moving on to the derivative formula in the next chapter is the following lemma on the asymptotics of derivatives. It will be stated for a scalar-valued function $w(\xi,\lambda,\nu)$, noting that it extends to vector-valued functions, e.g. $\mathbf{u}_j(\xi,\lambda,\nu)$ and $\mathbf{w}_j(\xi,\lambda,\nu)$, component-wise.

Lemma 6.3 *Suppose that*

$$\lim_{\xi\to+\infty} w(\xi,\lambda,\nu)e^{\mu(\lambda,\nu)\xi} = \eta(\lambda,\nu), \quad \xi\in\mathbb{R},\ (\lambda,\nu)\in\Lambda\,, \tag{6.46}$$

where w, η are analytic functions for $\lambda\in\Lambda$, the convergence is uniform for $(\lambda,\nu)\in\Lambda$, and $\mathrm{Re}(\mu(\lambda,\nu)) > 0$. Then the functions defined as

$$f_k(\xi,\lambda,\nu) := e^{\mu(\lambda,\nu)\xi}\frac{\partial^k}{\partial\lambda^k}w(\xi,\lambda,\nu)\,, \quad k\in\mathbb{N} \tag{6.47}$$

satisfy

$$f_k(\xi,\lambda,\nu) = O_{\xi\to+\infty}(\xi^k)\,, \tag{6.48}$$

for all $(\lambda,\nu)\in\Lambda$.

Proof The dependence on ν will be suppressed as it does not enter into the proof. First, define the function $g(\xi, \lambda)$ as

$$g(\xi, \lambda) := w(\xi, \lambda)e^{\mu(\lambda)\xi} - \eta(\lambda)\,. \tag{6.49}$$

This function is analytic for $\lambda \in \Lambda$ and goes to zero as $\xi \to \infty$ for all λ. This expression can be rewritten as

$$w(\xi, \lambda) = e^{-\mu(\lambda)\xi}[\eta(\lambda) + g(\xi, \lambda)]\,. \tag{6.50}$$

Differentiating with respect to λ gives us

$$w_\lambda(\xi, \lambda) = e^{-\mu(\lambda)\xi}\left[\eta_\lambda(\lambda) + g_\lambda(\xi, \lambda) - \mu_\lambda(\lambda)\xi[\eta(\lambda) + g(\xi, \lambda))]\right]. \tag{6.51}$$

From this we can see that

$$f_1(\xi, \lambda) = \eta_\lambda(\lambda) + g_\lambda(\xi, \lambda) - \mu_\lambda(\lambda)\xi[\eta(\lambda) + g(\xi, \lambda)]. \tag{6.52}$$

The highest-order term in f_1 for large ξ is $-\mu_\lambda(\lambda)\eta(\lambda)\xi$. Therefore, $f_1(\xi, \lambda) = O_{\xi\to+\infty}(\xi)$.

Now suppose that

$$f_j(\xi, \lambda) = O_{\xi\to+\infty}(\xi^j) \tag{6.53}$$

for some $j \in \mathbb{N}$. Then since

$$\frac{\partial^j}{\partial\lambda^j}w(\xi, \lambda) = e^{-\mu(\lambda)\xi}f_j(\xi, \lambda)\,, \tag{6.54}$$

we can differentiate with respect to λ to get

$$\frac{\partial^{(j+1)}}{\partial\lambda^{(j+1)}}w(\xi, \lambda) = e^{-\mu(\lambda)\xi}[(f_j)_\lambda(\xi, \lambda) - \mu_\lambda(\lambda)\xi f_j(\xi, \lambda)]\,. \tag{6.55}$$

Hence

$$f_{(j+1)}(\xi, \lambda) = (f_j)_\lambda(\xi, \lambda) - \mu_\lambda(\lambda)\xi f_j(\xi, \lambda)\,. \tag{6.56}$$

As $\xi \to \infty$ these two terms are respectively $O(\xi^j)$ and $O(\xi^{(j+1)})$. Therefore $f_{(j+1)}(\xi, \lambda) = O_{\xi\to+\infty}(\xi^{(j+1)})$ and so the proof of the lemma follows by induction. The same result holds for $\xi \to -\infty$ and $\mathrm{Re}(\mu(\lambda)) < 0$, and the proof strategy is identical. ∎

Chapter 7
Derivatives of the Evans Function

In this chapter, a proof of the leading-order terms in a Taylor expansion of $D(\lambda, \nu)$ is given, confirming the formulas introduced in the introduction (1.38). For reference, a summary of the hypotheses is given in Chap. 8.

Theorem 7.1 *The Evans function* (6.30)*, associated with the linearisation of the class of PDEs* (4.1)*, about the solitary wave solutions* (4.5)*, under the hypotheses* (**H1**)–(**H7**)*, has the following derivatives at* $\lambda = \nu = 0$*,*

$$
\begin{aligned}
&D(0,0) = 0\,, \quad D_\lambda(0,0) = 0\,, \quad D_\nu(0,0) = 0\,, \quad D_{\lambda\lambda}(0,0) = \chi\Pi\frac{\partial I_1}{\partial c}\,,\\
&D_{\lambda\nu}(0,0) = \mathrm{i}\chi\Pi\frac{\partial I_1}{\partial \ell} = \mathrm{i}\chi\Pi\frac{\partial I_2}{\partial c} \quad \text{and} \quad D_{\nu\nu}(0,0) = -\chi\Pi\frac{\partial I_2}{\partial \ell}\,,
\end{aligned}
$$

where Π *is the Lazutkin invariant* (5.21)*,* χ *is given in* (5.31)*, and* $I_1(c,\ell)$*,* $I_2(c,\ell)$ *are defined in* (4.9)*, evaluated on the* (c,ℓ)*-dependent family of solitary waves.*

Proof The proof begins by rewriting the matrix representation of $D(\lambda, \nu)$ in (6.30) as a block matrix

$$
D(\lambda,\nu) = \frac{1}{2}\det\begin{bmatrix} d_0(\lambda,\nu) & \mathbf{d}_1^T(\lambda,\nu) \\ \mathbf{d}_2(\lambda,\nu) & \mathcal{D}(\lambda,\nu) \end{bmatrix} \tag{7.1}
$$

with

$$
d_0 = \mathbf{\Omega}(\mathbf{w}_5,\mathbf{u}_5), \qquad \mathbf{d}_1 = \begin{pmatrix} \mathbf{\Omega}(\mathbf{w}_5,\mathbf{u}_6) \\ \mathbf{\Omega}(\mathbf{w}_5,\mathbf{u}_7) \\ \mathbf{\Omega}(\mathbf{w}_5,\mathbf{u}_8) \end{pmatrix},
$$

$$
\mathbf{d}_2 = \begin{pmatrix} \mathbf{\Omega}(\mathbf{w}_6,\mathbf{u}_5) \\ \mathbf{\Omega}(\mathbf{w}_7,\mathbf{u}_5) \\ \mathbf{\Omega}(\mathbf{w}_8,\mathbf{u}_5) \end{pmatrix}, \quad \mathcal{D} = \begin{pmatrix} \mathbf{\Omega}(\mathbf{w}_6,\mathbf{u}_6) & \mathbf{\Omega}(\mathbf{w}_6,\mathbf{u}_7) & \mathbf{\Omega}(\mathbf{w}_6,\mathbf{u}_8) \\ \mathbf{\Omega}(\mathbf{w}_7,\mathbf{u}_6) & \mathbf{\Omega}(\mathbf{w}_7,\mathbf{u}_7) & \mathbf{\Omega}(\mathbf{w}_7,\mathbf{u}_8) \\ \mathbf{\Omega}(\mathbf{w}_8,\mathbf{u}_6) & \mathbf{\Omega}(\mathbf{w}_8,\mathbf{u}_7) & \mathbf{\Omega}(\mathbf{w}_8,\mathbf{u}_8) \end{pmatrix}.
$$

T. J. Burchell, T. J. Bridges, *Transverse Instability of Solitary Waves*, Mathematical Physics Studies, https://doi.org/10.1007/978-981-95-2046-6_7

Now

$$\det[\mathcal{D}(0,0)] = \det \begin{pmatrix} \boldsymbol{\Omega}(\mathbf{a}^+, \mathbf{a}^-) & \boldsymbol{\Omega}(\mathbf{a}^+, \mathbf{b}^-) & \boldsymbol{\Omega}(\mathbf{a}^+, \mathbf{c}^-) \\ \boldsymbol{\Omega}(\mathbf{b}^+, \mathbf{a}^-) & \boldsymbol{\Omega}(\mathbf{b}^+, \mathbf{b}^-) & \boldsymbol{\Omega}(\mathbf{b}^+, \mathbf{c}^-) \\ \boldsymbol{\Omega}(\mathbf{c}^+, \mathbf{a}^-) & \boldsymbol{\Omega}(\mathbf{c}^+, \mathbf{b}^-) & \boldsymbol{\Omega}(\mathbf{c}^+, \mathbf{c}^-) \end{pmatrix} = -\Pi. \tag{7.2}$$

Since $\Pi \neq 0$ by hypothesis (**H5**), $\mathcal{D}$ must be invertible in a neighbourhood of the origin. This means that the Evans function matrix can be broken down into the following product:

$$\begin{pmatrix} d_0 & \mathbf{d}_1^T \\ \mathbf{d}_2 & \mathcal{D} \end{pmatrix} = \begin{pmatrix} 1 & \mathbf{d}_1^T \\ \mathbf{0} & \mathcal{D} \end{pmatrix} \begin{pmatrix} \gamma & \mathbf{0} \\ \mathcal{D}^{-1}\mathbf{d}_2 & \mathbf{I}_3 \end{pmatrix},$$

where $\gamma(\lambda, \nu) := d_0 - \mathbf{d}_1^T \mathcal{D}^{-1} \mathbf{d}_2$ is the Schur complement of $\mathcal{D}$. It is clear that the determinant of the first matrix on the RHS is $\det \mathcal{D}$, while that of the second is γ. Therefore, we find that:

$$D(\lambda, \nu) = \frac{1}{2} \gamma \det \mathcal{D}. \tag{7.3}$$

If we now bring in (6.44) and (6.45),

$$\begin{aligned} \mathbf{u}_5(\xi, 0, 0) &= (\chi^-)^{-1} \widehat{Z}_\xi, & \mathbf{w}_5(\xi, 0, 0) &= (\chi^+)^{-1} \widehat{Z}_\xi, \\ \mathbf{u}_6(\xi, 0, 0) &= \mathbf{a}^-, & \mathbf{w}_6(\xi, 0, 0) &= \mathbf{a}^+, \\ \mathbf{u}_7(\xi, 0, 0) &= \mathbf{b}^-, & \mathbf{w}_7(\xi, 0, 0) &= \mathbf{b}^+, \\ \mathbf{u}_8(\xi, 0, 0) &= \mathbf{c}^-, & \mathbf{w}_8(\xi, 0, 0) &= \mathbf{c}^+, \end{aligned}$$

we can use Proposition 5.1 to deduce that when $\lambda, \nu = 0$,

$$\boldsymbol{\Omega}(\mathbf{w}_5, \mathbf{u}_j) = 0 \quad \text{and} \quad \boldsymbol{\Omega}(\mathbf{w}_j, \mathbf{u}_5) = 0$$

for $j = 5, \ldots, 8$. Therefore, $d_0(0, 0) = 0$ and $\mathbf{d}_1(0, 0) = \mathbf{d}_2(0, 0) = \mathbf{0}$, which in turn means that $\gamma(0, 0) = 0$. Thus $D(0, 0) = 0$.

Next, the first derivatives at $\lambda, \nu = 0$ simplify to

$$D_\lambda(0,0) = -\frac{1}{2} \Pi \gamma_\lambda(0,0), \qquad D_\nu(0,0) = -\frac{1}{2} \Pi \gamma_\nu(0,0). \tag{7.4}$$

Since

$$\gamma_\lambda(\lambda, \nu) = (d_0)_\lambda - (\mathbf{d}_1)_\lambda^T \mathcal{D}^{-1} \mathbf{d}_2 - \mathbf{d}_1^T (\mathcal{D}^{-1})_\lambda \mathbf{d}_2 - \mathbf{d}_1^T \mathcal{D}^{-1} (\mathbf{d}_2)_\lambda,$$

we can see that

$$\gamma_\lambda(0,0) = (d_0)_\lambda(0,0),$$

and likewise

$$\gamma_\nu(0,0) = (d_0)_\nu(0,0)\,.$$

Now

$$(d_0)_\lambda(\lambda,\nu) = \mathbf{\Omega}(\partial_\lambda \mathbf{w}_5, \mathbf{u}_5) + \mathbf{\Omega}(\mathbf{w}_5, \partial_\lambda \mathbf{u}_5) \quad \text{and}$$
$$(d_0)_\nu(\lambda,\nu) = \mathbf{\Omega}(\partial_\nu \mathbf{w}_5, \mathbf{u}_5) + \mathbf{\Omega}(\mathbf{w}_5, \partial_\nu \mathbf{u}_5)\,. \tag{7.5}$$

Using the defining equations for $\mathbf{u}_5$, $\mathbf{w}_5$,

$$\mathbf{J}(\mathbf{u}_5)_\xi = [\mathbf{B} - \lambda\mathbf{M} - \mathtt{i}\nu\mathbf{N}]\mathbf{u}_5 \quad \text{and} \quad \mathbf{J}(\mathbf{w}_5)_\xi = [\mathbf{B} + \lambda\mathbf{M} + \mathtt{i}\nu\mathbf{N}]\mathbf{w}_5\,,$$

we can differentiate them with respect to λ, ν to get

$$\mathbf{J}(\mathbf{u}_5)_{\xi\lambda} = [\mathbf{B} - \lambda\mathbf{M} - \mathtt{i}\nu\mathbf{N}](\mathbf{u}_5)_\lambda - \mathbf{M}\mathbf{u}_5, \tag{7.6}$$
$$\mathbf{J}(\mathbf{w}_5)_{\xi\lambda} = [\mathbf{B} + \lambda\mathbf{M} + \mathtt{i}\nu\mathbf{N}](\mathbf{w}_5)_\lambda + \mathbf{M}\mathbf{w}_5, \tag{7.7}$$
$$\mathbf{J}(\mathbf{u}_5)_{\xi\nu} = [\mathbf{B} - \lambda\mathbf{M} - \mathtt{i}\nu\mathbf{N}](\mathbf{u}_5)_\nu - \mathbf{iN}\mathbf{u}_5, \tag{7.8}$$
$$\mathbf{J}(\mathbf{w}_5)_{\xi\nu} = [\mathbf{B} + \lambda\mathbf{M} + \mathtt{i}\nu\mathbf{N}](\mathbf{w}_5)_\nu + \mathbf{iN}\mathbf{w}_5. \tag{7.9}$$

Set $\lambda, \nu = 0$ to get

$$\mathbf{L}(\mathbf{u}_5)_\lambda\Big|_{\lambda,\nu=0} = (\chi^-)^{-1}\mathbf{M}\widehat{Z}_\xi\,, \tag{7.10}$$
$$\mathbf{L}(\mathbf{w}_5)_\lambda\Big|_{\lambda,\nu=0} = -(\chi^+)^{-1}\mathbf{M}\widehat{Z}_\xi\,, \tag{7.11}$$
$$\mathbf{L}(\mathbf{u}_5)_\nu\Big|_{\lambda,\nu=0} = \mathtt{i}(\chi^-)^{-1}\mathbf{N}\widehat{Z}_\xi\,, \tag{7.12}$$
$$\mathbf{L}(\mathbf{w}_5)_\nu\Big|_{\lambda,\nu=0} = -\mathtt{i}(\chi^+)^{-1}\mathbf{N}\widehat{Z}_\xi\,, \tag{7.13}$$

using (6.44) and (6.45). Now use Eqs. (5.7), giving

$$(\mathbf{u}_5)_\lambda\Big|_{\lambda,\nu=0} = (\chi^-)^{-1}\widehat{Z}_c + C_1\widehat{Z}_\xi\,, \tag{7.14}$$
$$(\mathbf{w}_5)_\lambda\Big|_{\lambda,\nu=0} = -(\chi^+)^{-1}\widehat{Z}_c + C_2\widehat{Z}_\xi\,, \tag{7.15}$$
$$(\mathbf{u}_5)_\nu\Big|_{\lambda,\nu=0} = \mathtt{i}(\chi^-)^{-1}\widehat{Z}_\ell + C_3\widehat{Z}_\xi\,, \tag{7.16}$$
$$(\mathbf{w}_5)_\nu\Big|_{\lambda,\nu=0} = -\mathtt{i}(\chi^+)^{-1}\widehat{Z}_\ell + C_4\widehat{Z}_\xi\,, \tag{7.17}$$

with C_j, $j = 1, \dots, 4$ arbitrary constants. Substitute these expressions into (7.5) evaluated at $\lambda, \nu = 0$,

$$
\begin{aligned}
(d_0)_\lambda(0,0) &= [\boldsymbol{\Omega}(\partial_\lambda \mathbf{w}_5, \mathbf{u}_5) + \boldsymbol{\Omega}(\mathbf{w}_5, \partial_\lambda \mathbf{u}_5)]\Big|_{\lambda,\nu=0} \\
&= \Big[\boldsymbol{\Omega}(-(\chi^+)^{-1}\widehat{Z}_c + C_2\widehat{Z}_\xi, (\chi^-)^{-1}\widehat{Z}_\xi) \\
&\quad + \boldsymbol{\Omega}((\chi^+)^{-1}\widehat{Z}_\xi, (\chi^-)^{-1}\widehat{Z}_c + C_1\widehat{Z}_\xi)\Big] \\
&= 2\chi\,\boldsymbol{\Omega}(\widehat{Z}_\xi, \widehat{Z}_c)\,, \qquad (7.18) \\
(d_0)_\nu(0,0) &= \Big[\boldsymbol{\Omega}(\partial_\nu \mathbf{w}_5, \mathbf{u}_5) + \boldsymbol{\Omega}(\mathbf{w}_5, \partial_\nu \mathbf{u}_5)\Big]\Big|_{\lambda,\nu=0} \\
&= \Big[\boldsymbol{\Omega}(-\mathtt{i}(\chi^+)^{-1}\widehat{Z}_\ell + C_4\widehat{Z}_\xi, (\chi^-)^{-1}\widehat{Z}_\xi) \\
&\quad + \boldsymbol{\Omega}((\chi^+)^{-1}\widehat{Z}_\xi, \mathtt{i}(\chi^-)^{-1}\widehat{Z}_\ell + C_3\widehat{Z}_\xi)\Big] \\
&= 2\mathtt{i}\chi\,\boldsymbol{\Omega}(\widehat{Z}_\xi, \widehat{Z}_\ell)\,. \qquad (7.19)
\end{aligned}
$$

Now since $\boldsymbol{\Omega}(\widehat{Z}_\xi, \widehat{Z}_c)$ and $\boldsymbol{\Omega}(\widehat{Z}_\xi, \widehat{Z}_\ell)$ both tend to zero as $\xi \to \pm\infty$ and

$$
\begin{aligned}
\frac{\mathrm{d}}{\mathrm{d}\xi}\boldsymbol{\Omega}(\widehat{Z}_\xi, \widehat{Z}_c) &= \langle \mathbf{J}(\widehat{Z}_\xi)_\xi, \widehat{Z}_c\rangle - \langle \widehat{Z}_\xi, \mathbf{J}(\widehat{Z}_c)_\xi\rangle \\
&= \langle \mathbf{B}\widehat{Z}_\xi, \widehat{Z}_c\rangle + \langle \widehat{Z}_\xi, \mathbf{M}\widehat{Z}_\xi\rangle - \langle \widehat{Z}_\xi, \mathbf{B}\widehat{Z}_c\rangle \\
&= 0, \\
\frac{\mathrm{d}}{\mathrm{d}\xi}\boldsymbol{\Omega}(\widehat{Z}_\xi, \widehat{Z}_\ell) &= \langle \mathbf{J}(\widehat{Z}_\xi)_\xi, \widehat{Z}_\ell\rangle - \langle \widehat{Z}_\xi, \mathbf{J}(\widehat{Z}_\ell)_\xi\rangle \\
&= \langle \mathbf{B}\widehat{Z}_\xi, \widehat{Z}_\ell\rangle + \langle \widehat{Z}_\xi, \mathbf{N}\widehat{Z}_\xi\rangle - \langle \widehat{Z}_\xi, \mathbf{B}\widehat{Z}_\ell\rangle \\
&= 0,
\end{aligned}
$$

they must be zero for all values of ξ so $\gamma_\lambda(0,0) = \gamma_\nu(0,0) = 0$ and thus $D_\lambda(0,0) = D_\nu(0,0) = 0$. It also allows us to simplify the second derivatives

$$
\begin{aligned}
D_{\lambda\lambda}(\lambda,\nu) &= \tfrac{1}{2}\left[\gamma_{\lambda\lambda}\det\mathcal{D} + 2\gamma_\lambda\,(\det\mathcal{D})_\lambda + \gamma\,(\det\mathcal{D})_{\lambda\lambda}\right], \\
D_{\lambda\nu}(\lambda,\nu) &= \tfrac{1}{2}\left[\gamma_{\lambda\nu}\det\mathcal{D} + \gamma_\lambda\,(\det\mathcal{D})_\nu + \gamma_\nu\,(\det\mathcal{D})_\lambda + \gamma\,(\det\mathcal{D})_{\lambda\nu}\right], \\
D_{\nu\nu}(\lambda,\nu) &= \tfrac{1}{2}\left[\gamma_{\nu\nu}\det\mathcal{D} + 2\gamma_\nu\,(\det\mathcal{D})_\nu + \gamma\,(\det\mathcal{D})_{\nu\nu}\right],
\end{aligned}
$$

at $\lambda = 0$, $\nu = 0$ to get:

$$
D_{\lambda\lambda}(0,0) = -\frac{1}{2}\Pi\gamma_{\lambda\lambda}(0,0),
$$

$$\begin{aligned} D_{\lambda\nu}(0,0) &= -\frac{1}{2}\Pi\gamma_{\lambda\nu}(0,0), \\ D_{\nu\nu}(0,0) &= -\frac{1}{2}\Pi\gamma_{\nu\nu}(0,0). \end{aligned} \tag{7.20}$$

Now by using the fact that $\mathbf{d}_1(0,0) = \mathbf{d}_2(0,0) = \mathbf{0}$ we can see that the second derivatives of $\gamma(0,0)$ simplify to:

$$\begin{aligned} \gamma_{\lambda\lambda}(0,0) &= (d_0)_{\lambda\lambda} - 2(\mathbf{d}_1)_\lambda^T \mathcal{D}^{-1}(\mathbf{d}_2)_\lambda, \\ \gamma_{\lambda\nu}(0,0) &= (d_0)_{\lambda\nu} - (\mathbf{d}_1)_\lambda^T \mathcal{D}^{-1}(\mathbf{d}_2)_\nu - (\mathbf{d}_1)_\nu^T \mathcal{D}^{-1}(\mathbf{d}_2)_\lambda, \\ \gamma_{\nu\nu}(0,0) &= (d_0)_{\nu\nu} - 2(\mathbf{d}_1)_\nu^T \mathcal{D}^{-1}(\mathbf{d}_2)_\nu. \end{aligned}$$

If we consider the components of $\mathbf{d}_1$ and take derivatives w.r.t λ and ν at $\lambda = 0$, $\nu = 0$ we get:

$$\begin{aligned} \partial_\lambda \boldsymbol{\Omega}(\mathbf{w}_5(\xi,0,0), \mathbf{u}_j(\xi,0,0)) &= \boldsymbol{\Omega}(-(\chi^+)^{-1}\widehat{Z}_c + C_2\widehat{Z}_\xi, \mathbf{u}_j(\xi,0,0)) \\ &\quad + \boldsymbol{\Omega}((\chi^+)^{-1}\widehat{Z}_\xi, (\mathbf{u}_j)_\lambda(\xi,0,0)), \\ \partial_\nu \boldsymbol{\Omega}(\mathbf{w}_5(\xi,0,0), \mathbf{u}_j(\xi,0,0)) &= \boldsymbol{\Omega}(-\mathrm{i}(\chi^+)^{-1}\widehat{Z}_\ell + C_4\widehat{Z}_\xi, \mathbf{u}_j(\xi,0,0)) \\ &\quad + \boldsymbol{\Omega}((\chi^+)^{-1}\widehat{Z}_\xi, (\mathbf{u}_j)_\nu(\xi,0,0)) \end{aligned}$$

for $j = 6, 7, 8$. Since these expressions are independent of ξ and can be shown to go to zero as $\xi \to -\infty$ (via Lemma 6.3) we can conclude that $(\mathbf{d}_1)_\lambda(0,0) = (\mathbf{d}_1)_\nu(0,0) = \mathbf{0}$. This means that (7.20) can be rewritten as:

$$\begin{aligned} D_{\lambda\lambda}(0,0) &= -\frac{1}{2}\Pi\,(d_0)_{\lambda\lambda}(0,0)\,, \\ D_{\lambda\nu}(0,0) &= -\frac{1}{2}\Pi\,(d_0)_{\lambda\nu}(0,0)\,, \\ D_{\nu\nu}(0,0) &= -\frac{1}{2}\Pi\,(d_0)_{\nu\nu}(0,0)\,. \end{aligned} \tag{7.21}$$

To work out the second derivatives of d_0 we can use Eqs. (7.6)–(7.9) to write

$$\begin{aligned} \partial_\xi\langle \mathbf{J}(\mathbf{w}_5)_\lambda, \mathbf{u}_5\rangle &= \langle[\mathbf{B}+\lambda\mathbf{M}+\mathrm{i}\nu\mathbf{N}](\mathbf{w}_5)_\lambda, \mathbf{u}_5\rangle + \langle\mathbf{M}\mathbf{w}_5, \mathbf{u}_5\rangle \\ &\quad - \langle(\mathbf{w}_5)_\lambda, [\mathbf{B}-\lambda\mathbf{M}-\mathrm{i}\nu\mathbf{N}]\mathbf{u}_5\rangle, \end{aligned} \tag{7.22}$$

$$\begin{aligned} \partial_\xi\langle \mathbf{J}\mathbf{w}_5, (\mathbf{u}_5)_\lambda\rangle &= \langle[\mathbf{B}+\lambda\mathbf{M}+\mathrm{i}\nu\mathbf{N}]\mathbf{w}_5, (\mathbf{u}_5)_\lambda\rangle - \langle\mathbf{w}_5, [\mathbf{B}-\lambda\mathbf{M}-\mathrm{i}\nu\mathbf{N}](\mathbf{u}_5)_\lambda\rangle \\ &\quad + \langle\mathbf{w}_5, \mathbf{M}\mathbf{u}_5\rangle, \end{aligned} \tag{7.23}$$

$$\begin{aligned} \partial_\xi\langle \mathbf{J}(\mathbf{w}_5)_\nu, \mathbf{u}_5\rangle &= \langle[\mathbf{B}+\lambda\mathbf{M}+\mathrm{i}\nu\mathbf{N}](\mathbf{w}_5)_\nu, \mathbf{u}_5\rangle + \mathrm{i}\langle\mathbf{N}\mathbf{w}_5, \mathbf{u}_5\rangle \\ &\quad - \langle(\mathbf{w}_5)_\nu, [\mathbf{B}-\lambda\mathbf{M}-\mathrm{i}\nu\mathbf{N}]\mathbf{u}_5\rangle, \end{aligned} \tag{7.24}$$

$$\partial_\xi \langle \mathbf{J}\mathbf{w}_5, (\mathbf{u}_5)_\nu \rangle = \langle [\mathbf{B} + \lambda \mathbf{M} + \mathrm{i}\nu \mathbf{N}]\mathbf{w}_5, (\mathbf{u}_5)_\nu \rangle - \langle \mathbf{w}_5, [\mathbf{B} - \lambda \mathbf{M} - \mathrm{i}\nu \mathbf{N}](\mathbf{u}_5)_\nu \rangle \\ + \mathrm{i}\langle \mathbf{w}_5, \mathbf{N}\mathbf{u}_5 \rangle, \tag{7.25}$$

which leaves us with:

$$\partial_\xi \mathbf{\Omega}\left((\mathbf{w}_5)_\lambda, \mathbf{u}_5\right) = \langle \mathbf{M}\mathbf{w}_5, \mathbf{u}_5 \rangle = -\partial_\xi \mathbf{\Omega}\left(\mathbf{w}_5, (\mathbf{u}_5)_\lambda\right), \tag{7.26}$$

$$\partial_\xi \mathbf{\Omega}\left((\mathbf{w}_5)_\nu, \mathbf{u}_5\right) = \mathrm{i}\langle \mathbf{N}\mathbf{w}_5, \mathbf{u}_5 \rangle = -\partial_\xi \mathbf{\Omega}\left(\mathbf{w}_5, (\mathbf{u}_5)_\nu\right). \tag{7.27}$$

If we now take some $R > 0$ then we can integrate the first part of each of these over the range $\xi \in [0, R]$ and the second part of each over $\xi \in [-R, 0]$ to get:

$$\mathbf{\Omega}\left((\mathbf{w}_5)_\lambda, \mathbf{u}_5\right)\Big|_{\xi=0}^{\xi=R} = \int_0^R \langle \mathbf{M}\mathbf{w}_5, \mathbf{u}_5 \rangle \, \mathrm{d}\xi, \tag{7.28}$$

$$-\mathbf{\Omega}\left(\mathbf{w}_5, (\mathbf{u}_5)_\lambda\right)\Big|_{\xi=-R}^{\xi=0} = \int_{-R}^0 \langle \mathbf{M}\mathbf{w}_5, \mathbf{u}_5 \rangle \, \mathrm{d}\xi, \tag{7.29}$$

$$\mathbf{\Omega}\left((\mathbf{w}_5)_\nu, \mathbf{u}_5\right)\Big|_{\xi=0}^{\xi=R} = \mathrm{i}\int_0^R \langle \mathbf{N}\mathbf{w}_5, \mathbf{u}_5 \rangle \, \mathrm{d}\xi, \tag{7.30}$$

$$-\mathbf{\Omega}\left(\mathbf{w}_5, (\mathbf{u}_5)_\nu\right)\Big|_{\xi=-R}^{\xi=0} = \mathrm{i}\int_{-R}^0 \langle \mathbf{N}\mathbf{w}_5, \mathbf{u}_5 \rangle \, \mathrm{d}\xi, \tag{7.31}$$

which combine to give:

$$(d_0)_\lambda(\lambda, \nu) = -\int_{-R}^R \langle \mathbf{M}\mathbf{w}_5, \mathbf{u}_5 \rangle \, \mathrm{d}\xi + \ell_R^{(1)}(\lambda, \nu), \tag{7.32}$$

$$(d_0)_\nu(\lambda, \nu) = -\mathrm{i}\int_{-R}^R \langle \mathbf{N}\mathbf{w}_5, \mathbf{u}_5 \rangle \, \mathrm{d}\xi + \ell_R^{(2)}(\lambda, \nu), \tag{7.33}$$

where the functions $\ell_R^{(1)}, \ell_R^{(2)}$ are defined as

$$\ell_R^{(1)}(\lambda, \nu) = \mathbf{\Omega}\left((\mathbf{w}_5)_\lambda, \mathbf{u}_5\right)\Big|_{\xi=R} + \mathbf{\Omega}\left(\mathbf{w}_5, (\mathbf{u}_5)_\lambda\right)\Big|_{\xi=-R},$$

$$\ell_R^{(2)}(\lambda, \nu) = \mathbf{\Omega}\left((\mathbf{w}_5)_\nu, \mathbf{u}_5\right)\Big|_{\xi=R} + \mathbf{\Omega}\left(\mathbf{w}_5, (\mathbf{u}_5)_\nu\right)\Big|_{\xi=-R}.$$

Differentiating with respect to λ and ν gives us expressions for $(d_0)_{\lambda\lambda}(\lambda, \nu)$ and $(d_0)_{\nu\nu}(\lambda, \nu)$,

$$(d_0)_{\lambda\lambda}(\lambda, \nu) = -\int_{-R}^R \langle \mathbf{M}(\mathbf{w}_5)_\lambda, \mathbf{u}_5 \rangle + \langle \mathbf{M}\mathbf{w}_5, (\mathbf{u}_5)_\lambda \rangle \, \mathrm{d}\xi + (\ell_R^{(1)})_\lambda(\lambda, \nu), \tag{7.34}$$

$$(d_0)_{\nu\nu}(\lambda,\nu) = -\mathrm{i}\int_{-R}^{R} \langle \mathbf{N}(\mathbf{w}_5)_\nu, \mathbf{u}_5\rangle + \langle \mathbf{N}\mathbf{w}_5, (\mathbf{u}_5)_\nu\rangle \, \mathrm{d}\xi + (\ell_R^{(2)})_\nu(\lambda,\nu), \quad (7.35)$$

along with two expressions for $(d_0)_{\lambda\nu}(\lambda,\nu)$,

$$\begin{aligned}(d_0)_{\lambda\nu}(\lambda,\nu) &= -\int_{-R}^{R} \langle \mathbf{M}(\mathbf{w}_5)_\nu, \mathbf{u}_5\rangle + \langle \mathbf{M}\mathbf{w}_5, (\mathbf{u}_5)_\nu\rangle \, \mathrm{d}\xi + (\ell_R^{(1)})_\nu(\lambda,\nu)\,,\\ &= -\mathrm{i}\int_{-R}^{R} \langle \mathbf{N}(\mathbf{w}_5)_\lambda, \mathbf{u}_5\rangle + \langle \mathbf{N}\mathbf{w}_5, (\mathbf{u}_5)_\lambda\rangle \, \mathrm{d}\xi + (\ell_R^{(2)})_\lambda(\lambda,\nu)\,.\end{aligned} \quad (7.36)$$

Taking the limit $R \to \infty$ allows us to write these expressions as

$$(d_0)_{\lambda\lambda}(\lambda,\nu) = -\int_{-\infty}^{+\infty} \langle \mathbf{M}(\mathbf{w}_5)_\lambda, \mathbf{u}_5\rangle + \langle \mathbf{M}\mathbf{w}_5, (\mathbf{u}_5)_\lambda\rangle \, \mathrm{d}\xi + (\ell_\infty^{(1)})_\lambda(\lambda,\nu), \quad (7.37)$$

$$(d_0)_{\nu\nu}(\lambda,\nu) = -\mathrm{i}\int_{-\infty}^{+\infty} \langle \mathbf{N}(\mathbf{w}_5)_\nu, \mathbf{u}_5\rangle + \langle \mathbf{N}\mathbf{w}_5, (\mathbf{u}_5)_\nu\rangle \, \mathrm{d}\xi + (\ell_\infty^{(2)})_\nu(\lambda,\nu) \quad (7.38)$$

and

$$\begin{aligned}(d_0)_{\lambda\nu}(\lambda,\nu) &= -\int_{-\infty}^{+\infty} \langle \mathbf{M}(\mathbf{w}_5)_\nu, \mathbf{u}_5\rangle + \langle \mathbf{M}\mathbf{w}_5, (\mathbf{u}_5)_\nu\rangle \, \mathrm{d}\xi + (\ell_\infty^{(1)})_\nu(\lambda,\nu)\,,\\ &= -\mathrm{i}\int_{-\infty}^{+\infty} \langle \mathbf{N}(\mathbf{w}_5)_\lambda, \mathbf{u}_5\rangle + \langle \mathbf{N}\mathbf{w}_5, (\mathbf{u}_5)_\lambda\rangle \, \mathrm{d}\xi + (\ell_\infty^{(2)})_\lambda(\lambda,\nu)\,,\end{aligned} \quad (7.39)$$

where

$$\begin{aligned}(\ell_\infty^{(1)})_\lambda(\lambda,\nu) &:= \lim_{\xi\to+\infty} \left[\boldsymbol{\Omega}\left((\mathbf{w}_5)_{\lambda\lambda}, \mathbf{u}_5\right) + \boldsymbol{\Omega}\left((\mathbf{w}_5)_\lambda, (\mathbf{u}_5)_\lambda\right)\right]\\ &\quad + \lim_{\xi\to-\infty} \left[\boldsymbol{\Omega}\left((\mathbf{w}_5)_\lambda, (\mathbf{u}_5)_\lambda\right) + \boldsymbol{\Omega}\left(\mathbf{w}_5, (\mathbf{u}_5)_{\lambda\lambda}\right)\right],\\ (\ell_\infty^{(1)})_\nu(\lambda,\nu) &:= \lim_{\xi\to+\infty} \left[\boldsymbol{\Omega}\left((\mathbf{w}_5)_{\lambda\nu}, \mathbf{u}_5\right) + \boldsymbol{\Omega}\left((\mathbf{w}_5)_\lambda, (\mathbf{u}_5)_\nu\right)\right]\\ &\quad + \lim_{\xi\to-\infty} \left[\boldsymbol{\Omega}\left((\mathbf{w}_5)_\nu, (\mathbf{u}_5)_\lambda\right) + \boldsymbol{\Omega}\left(\mathbf{w}_5, (\mathbf{u}_5)_{\lambda\nu}\right)\right],\\ (\ell_\infty^{(2)})_\lambda(\lambda,\nu) &:= \lim_{\xi\to+\infty} \left[\boldsymbol{\Omega}\left((\mathbf{w}_5)_{\lambda\nu}, \mathbf{u}_5\right) + \boldsymbol{\Omega}\left((\mathbf{w}_5)_\nu, (\mathbf{u}_5)_\lambda\right)\right]\\ &\quad + \lim_{\xi\to-\infty} \left[\boldsymbol{\Omega}\left((\mathbf{w}_5)_\lambda, (\mathbf{u}_5)_\nu\right) + \boldsymbol{\Omega}\left(\mathbf{w}_5, (\mathbf{u}_5)_{\lambda\nu}\right)\right],\\ (\ell_\infty^{(2)})_\nu(\lambda,\nu) &:= \lim_{\xi\to+\infty} \left[\boldsymbol{\Omega}\left((\mathbf{w}_5)_{\nu\nu}, \mathbf{u}_5\right) + \boldsymbol{\Omega}\left((\mathbf{w}_5)_\nu, (\mathbf{u}_5)_\nu\right)\right]\\ &\quad + \lim_{\xi\to-\infty} \left[\boldsymbol{\Omega}\left((\mathbf{w}_5)_\nu, (\mathbf{u}_5)_\nu\right) + \boldsymbol{\Omega}\left(\mathbf{w}_5, (\mathbf{u}_5)_{\nu\nu}\right)\right].\end{aligned}$$

Since

$$\lim_{\xi\to+\infty} e^{\mu_5(\lambda,\nu)\xi}\mathbf{w}_5 = \eta_5(\lambda,\nu)$$

we can apply Lemma 6.3 to get

$$\begin{aligned} e^{\mu_5\xi}(\mathbf{w}_5)_{\lambda\lambda} &= O_{\xi\to+\infty}(\xi^2)\,, \\ e^{\mu_5\xi}(\mathbf{w}_5)_{\lambda\nu} &= O_{\xi\to+\infty}(\xi^2)\,, \\ e^{\mu_5\xi}(\mathbf{w}_5)_{\nu\nu} &= O_{\xi\to+\infty}(\xi^2)\,. \end{aligned}$$

Since the exponential will dominate the polynomial in each case this tells us that

$$\lim_{\xi\to+\infty}(\mathbf{w}_5)_{\lambda\lambda} = \lim_{\xi\to+\infty}(\mathbf{w}_5)_{\lambda\nu} = \lim_{\xi\to+\infty}(\mathbf{w}_5)_{\nu\nu} = 0.$$

The same argument can be used to show that

$$\lim_{\xi\to-\infty}(\mathbf{u}_5)_{\lambda\lambda} = \lim_{\xi\to-\infty}(\mathbf{u}_5)_{\lambda\nu} = \lim_{\xi\to-\infty}(\mathbf{u}_5)_{\nu\nu} = 0.$$

Now

$$\begin{aligned} (\ell_\infty^{(1)})_\lambda(0,0) = &\lim_{\xi\to+\infty}\Big[\boldsymbol{\Omega}\Big((\mathbf{w}_5)_{\lambda\lambda}|_{\lambda,\nu=0}\,,(\chi^-)^{-1}\widehat{Z}_\xi\Big) \\ &+\boldsymbol{\Omega}\Big(-(\chi^+)^{-1}\widehat{Z}_c + C_2\widehat{Z}_\xi,(\chi^-)^{-1}\widehat{Z}_c + C_1\widehat{Z}_\xi\Big)\Big] \\ &+\lim_{\xi\to-\infty}\Big[\boldsymbol{\Omega}\Big(-(\chi^+)^{-1}\widehat{Z}_c + C_2\widehat{Z}_\xi,(\chi^-)^{-1}\widehat{Z}_c + C_1\widehat{Z}_\xi\Big) \\ &+\boldsymbol{\Omega}\Big((\chi^+)^{-1}\widehat{Z}_\xi,\,(\mathbf{u}_5)_{\lambda\lambda}|_{\lambda,\nu=0}\Big)\Big], \\ (\ell_\infty^{(1)})_\nu(0,0) = &\lim_{\xi\to+\infty}\Big[\boldsymbol{\Omega}\Big((\mathbf{w}_5)_{\lambda\nu}|_{\lambda,\nu=0}\,,(\chi^-)^{-1}\widehat{Z}_\xi\Big) \\ &+\boldsymbol{\Omega}\Big(-(\chi^+)^{-1}\widehat{Z}_c + C_2\widehat{Z}_\xi,\mathrm{i}(\chi^-)^{-1}\widehat{Z}_\ell + C_3\widehat{Z}_\xi\Big)\Big] \\ &+\lim_{\xi\to-\infty}\Big[\boldsymbol{\Omega}\Big(-\mathrm{i}(\chi^+)^{-1}\widehat{Z}_\ell + C_4\widehat{Z}_\xi,(\chi^-)^{-1}\widehat{Z}_c + C_1\widehat{Z}_\xi\Big) \\ &+\boldsymbol{\Omega}\Big((\chi^+)^{-1}\widehat{Z}_\xi,\,(\mathbf{u}_5)_{\lambda\nu}|_{\lambda,\nu=0}\Big)\Big], \end{aligned}$$

$$\begin{aligned} (\ell_\infty^{(2)})_\lambda(0,0) = &\lim_{\xi\to+\infty}\Big[\boldsymbol{\Omega}\Big((\mathbf{w}_5)_{\lambda\nu}|_{\lambda,\nu=0}\,,(\chi^-)^{-1}\widehat{Z}_\xi\Big) \\ &+\boldsymbol{\Omega}\Big(-\mathrm{i}(\chi^+)^{-1}\widehat{Z}_\ell + C_4\widehat{Z}_\xi,(\chi^-)^{-1}\widehat{Z}_c + C_1\widehat{Z}_\xi\Big)\Big] \end{aligned}$$

$$+\lim_{\xi\to-\infty}\Big[\boldsymbol{\Omega}\left(-(\chi^+)^{-1}\widehat{Z}_c + C_2\widehat{Z}_\xi, \mathrm{i}(\chi^-)^{-1}\widehat{Z}_\ell + C_3\widehat{Z}_\xi\right)$$
$$+\boldsymbol{\Omega}\left((\chi^+)^{-1}\widehat{Z}_\xi, (\mathbf{u}_5)_{\lambda\nu}|_{\lambda,\nu=0}\right)\Big],$$
$$(\ell^{(2)}_\infty)_\nu(0,0) = \lim_{\xi\to+\infty}\Big[\boldsymbol{\Omega}\left((\mathbf{w}_5)_{\nu\nu}|_{\lambda,\nu=0}, (\chi^-)^{-1}\widehat{Z}_\xi\right)$$
$$+\boldsymbol{\Omega}\left(-\mathrm{i}(\chi^+)^{-1}\widehat{Z}_\ell + C_4\widehat{Z}_\xi, \mathrm{i}(\chi^-)^{-1}\widehat{Z}_\ell + C_3\widehat{Z}_\xi\right)\Big]$$
$$+\lim_{\xi\to-\infty}\Big[\boldsymbol{\Omega}\left(-\mathrm{i}(\chi^+)^{-1}\widehat{Z}_\ell + C_4\widehat{Z}_\xi, \mathrm{i}(\chi^-)^{-1}\widehat{Z}_\ell + C_3\widehat{Z}_\xi\right)$$
$$+\boldsymbol{\Omega}\left((\chi^+)^{-1}\widehat{Z}_\xi, (\mathbf{u}_5)_{\nu\nu}|_{\lambda,\nu=0}\right)\Big].$$

However, it is clear that every term in these expressions will go to zero. Therefore

$$(\ell^{(1)}_\infty)_\lambda(0,0) = (\ell^{(1)}_\infty)_\nu(0,0) = (\ell^{(2)}_\infty)_\lambda(0,0) = (\ell^{(2)}_\infty)_\nu(0,0) = 0,$$

leaving us with

$$(d_0)_{\lambda\lambda}(0,0) = -\int_{-\infty}^{+\infty}\langle\mathbf{M}(-(\chi^+)^{-1}\widehat{Z}_c + C_2\widehat{Z}_\xi), (\chi^-)^{-1}\widehat{Z}_\xi\rangle$$
$$+\langle(\chi^+)^{-1}\mathbf{M}\widehat{Z}_\xi, (\chi^-)^{-1}\widehat{Z}_c + C_1\widehat{Z}_\xi\rangle\,\mathrm{d}\xi$$
$$= -2\chi\int_{-\infty}^{+\infty}\langle\mathbf{M}\widehat{Z}_\xi, \widehat{Z}_c\rangle\,\mathrm{d}\xi$$
$$= -2\chi\frac{\partial I_1}{\partial c}, \tag{7.40}$$
$$(d_0)_{\nu\nu}(0,0) = -\mathrm{i}\int_{-\infty}^{+\infty}\langle\mathbf{N}(-\mathrm{i}(\chi^+)^{-1}\widehat{Z}_\ell + C_4\widehat{Z}_\xi), (\chi^-)^{-1}\widehat{Z}_\xi\rangle$$
$$+\langle(\chi^+)^{-1}\mathbf{N}\widehat{Z}_\xi, \mathrm{i}(\chi^-)^{-1}\widehat{Z}_\ell + C_3\widehat{Z}_\xi\rangle\,\mathrm{d}\xi$$
$$= 2\chi\int_{-\infty}^{+\infty}\langle\mathbf{N}\widehat{Z}_\xi, \widehat{Z}_\ell\rangle\,\mathrm{d}\xi$$
$$= 2\chi\frac{\partial I_2}{\partial \ell}, \tag{7.41}$$

and

$$(d_0)_{\lambda\nu}(0,0) = -\int_{-\infty}^{+\infty}\langle\mathbf{M}(-\mathrm{i}(\chi^+)^{-1}\widehat{Z}_\ell + C_4\widehat{Z}_\xi), (\chi^-)^{-1}\widehat{Z}_\xi\rangle$$
$$+\langle(\chi^+)^{-1}\mathbf{M}\widehat{Z}_\xi, \mathrm{i}(\chi^-)^{-1}\widehat{Z}_\ell + C_3\widehat{Z}_\xi\rangle\,\mathrm{d}\xi$$

$$
\begin{aligned}
&= -2\mathrm{i}\chi \int_{-\infty}^{+\infty} \langle \mathbf{M}\widehat{Z}_\xi, \widehat{Z}_\ell \rangle \, \mathrm{d}\xi \\
&= -2\mathrm{i}\chi \frac{\partial I_1}{\partial \ell}, \\
\text{or} \quad &= -\mathrm{i} \int_{-\infty}^{+\infty} \langle \mathbf{N}(-(\chi^+)^{-1}\widehat{Z}_c + C_2\widehat{Z}_\xi), (\chi^-)^{-1}\widehat{Z}_\xi \rangle \\
&\quad + \langle (\chi^+)^{-1}\mathbf{N}\widehat{Z}_\xi, (\chi^-)^{-1}\widehat{Z}_c + C_1\widehat{Z}_\xi \rangle \, \mathrm{d}\xi \\
&= -2\mathrm{i}\chi \int_{-\infty}^{+\infty} \langle \mathbf{N}\widehat{Z}_\xi, \widehat{Z}_c \rangle \, \mathrm{d}\xi \\
&= -2\mathrm{i}\chi \frac{\partial I_2}{\partial c}.
\end{aligned}
\tag{7.42}
$$

This then gives us the final result

$$
\begin{aligned}
D_{\lambda\lambda}(0,0) &= \chi \Pi \frac{\partial I_1}{\partial c}, \\
D_{\lambda\nu}(0,0) &= \mathrm{i}\chi \Pi \frac{\partial I_1}{\partial \ell} = \mathrm{i}\chi \Pi \frac{\partial I_2}{\partial c}, \\
D_{\nu\nu}(0,0) &= -\chi \Pi \frac{\partial I_2}{\partial \ell},
\end{aligned}
\tag{7.43}
$$

thus proving the theorem. ∎

Corollary 7.2 *When the determinant*

$$
\det \begin{bmatrix} \dfrac{\partial I_1}{\partial c} & \dfrac{\partial I_1}{\partial \ell} \\ \dfrac{\partial I_2}{\partial c} & \dfrac{\partial I_2}{\partial \ell} \end{bmatrix}
$$

is positive, there exists a real unstable eigenvalue when $|\lambda|^2 + \nu^2$ *is sufficiently small.*

Proof With the derivatives at $\lambda = \nu = 0$ given in (7.1), the Evans function has the leading-order Taylor expansion

$$
D(\lambda, \nu) = \frac{1}{2}\Pi\chi \left(\frac{\partial I_1}{\partial c}\lambda^2 + \mathrm{i}\nu\lambda \left(\frac{\partial I_1}{\partial \ell} + \frac{\partial I_2}{\partial c} \right) + (\mathrm{i}\nu)^2 \frac{\partial I_2}{\partial \ell} \right) + o(|\lambda^2| + \nu^2).
$$

The term in brackets factors into

$$
\frac{\partial I_1}{\partial c} \left(\lambda + \mathrm{i}\nu \frac{\frac{\partial I_1}{\partial \ell}}{\frac{\partial I_1}{\partial c}} \right)^2 + \frac{1}{\frac{\partial I_1}{\partial c}} \left(\left(\frac{\partial I_1}{\partial \ell} \right)^2 - \frac{\partial I_2}{\partial c} \frac{\partial I_1}{\partial \ell} \right) \nu^2 .
$$

Setting this expression to zero gives

$$\left(\lambda + \mathrm{i}\nu \frac{\frac{\partial I_1}{\partial \ell}}{\frac{\partial I_1}{\partial c}}\right)^2 = \frac{\nu^2}{\left(\frac{\partial I_1}{\partial c}\right)^2} \det \begin{bmatrix} \partial_c I_1 & \partial_\ell I_1 \\ \partial_c I_2 & \partial_\ell I_2 \end{bmatrix}.$$

Hence, a real eigenvalue λ with positive real part exists when the determinant condition (1.40) is satisfied, for $|\lambda|^2 + \nu^2$ sufficiently small. ■

The factor Π in the derivative formula (7.1) is connected to the parity of the Maslov index via the formula (1.39). This connection is discussed further in Appendix D.

Chapter 8
Summary of Hypotheses Used

In this chapter, we summarise the hypotheses.
(**H1**)—The class of PDEs for which the theory is developed is PDEs of the form

$$\mathbf{M}Z_t + \mathbf{K}Z_x + \mathbf{N}Z_y = \nabla S(Z)\,, \quad Z \in \mathbb{R}^8\,,$$

where $S : \mathbb{R}^8 \to \mathbb{R}$ is taken to be a smooth scalar-valued function, ∇ the gradient on $\mathbb{R}^8$, and $\mathbf{M}, \mathbf{K}, \mathbf{N}$ are 8×8 constant skew-symmetric matrices which are non-degenerate in the following combination,

$$\det\big(\mathbf{J}(c,\ell)\big) \neq 0\,, \quad \text{where } \mathbf{J}(c,\ell) := \mathbf{K} + c\mathbf{M} + \ell\mathbf{N}\,, \quad \text{for all } (c,\ell) \in \mathscr{C}$$

for some non-empty open set $\mathscr{C} \subset \mathbb{R}^2$. The set $\mathscr{C}$ is chosen as the set of all (c,ℓ) for which a solitary wave exists.
(**H2**)—There exists a solitary wave solution $\widehat{Z}(\xi)$, where

$$\xi = x + \ell y + ct + \xi_0\,, \quad \text{and} \quad \lim_{\xi\to\pm\infty} \|\widehat{Z}(\xi)\| = 0\,,$$

which is a solution to

$$\mathbf{J}(c,\ell)\widehat{Z}_\xi = \nabla S(\widehat{Z})$$

and is a smooth function of ξ, c and ℓ for all $(c,\ell) \in \mathscr{C}$. In addition, there exist functionals

$$I_1(\widehat{Z}) := \int_{-\infty}^{+\infty} \frac{1}{2}\langle \mathbf{M}\widehat{Z}_\xi, \widehat{Z}\rangle \,\mathrm{d}\xi\,, \quad \text{and} \quad I_2(\widehat{Z}) = \int_{-\infty}^{+\infty} \frac{1}{2}\langle \mathbf{N}\widehat{Z}_\xi, \widehat{Z}\rangle \,\mathrm{d}\xi\,,$$

T. J. Burchell, T. J. Bridges, *Transverse Instability of Solitary Waves*, Mathematical Physics Studies, https://doi.org/10.1007/978-981-95-2046-6_8

which satisfy

$$\det\begin{bmatrix} \dfrac{\partial I_1}{\partial c} & \dfrac{\partial I_1}{\partial \ell} \\ \dfrac{\partial I_2}{\partial c} & \dfrac{\partial I_2}{\partial \ell} \end{bmatrix} \neq 0,$$

when evaluated on the family of solitary waves.
(H3)—The kernel of the linear operator

$$\mathbf{L} = \mathbf{B}(\xi, c, \ell) - \mathbf{J}(c, \ell)\partial_\xi\,, \quad \mathbf{B}(\xi, c, \ell) := D^2 S(\widehat{Z}(\xi, c, \ell))$$

is subject to the following condition:

$$\mathrm{Ker}(\mathbf{L}) \cap \left(\mathbb{R}^8 \otimes L^2(\mathbb{R})\right) = \mathrm{span}\{\widehat{Z}_\xi\}\,.$$

(H4)—The eigenvalues at infinity $\mu_j(0, 0, c, \ell)$, determined by

$$\Delta(\mu, 0, 0; c, \ell) := \det\left[\mathbf{B}^\infty(c, \ell) - \mu\mathbf{J}(c, \ell)\right] = 0\,,$$

are simple, real, and form a 4–4 splitting symmetric about the origin for all $(c, \ell) \in \mathscr{C}$. The eigenvectors ζ_j and adjoint eigenvectors η_i, satisfying

$$\begin{aligned} &\left[\mathbf{B}^\infty(c, \ell) - \mu_j\mathbf{J}(c, \ell)\right]\zeta_j = 0 \quad \text{and} \\ &\left[\mathbf{B}^\infty(c, \ell) + \mu_i\mathbf{J}(c, \ell)\right]\eta_j = 0\,, \quad j = 1, \ldots, 8\,, \end{aligned}$$

are normalised by setting

$$\mathbf{\Omega}(\eta_i, \zeta_j) := \langle \mathbf{J}(c, \ell)\eta_i, \zeta_j\rangle = \delta_{ij}\,, \quad i, j = 1, \ldots, 8\,.$$

(H5)—The Lazutkin–Treschev invariant takes the form

$$\Pi := \det\begin{bmatrix} \mathbf{\Omega}(\mathbf{a}^-, \mathbf{a}^+) & \mathbf{\Omega}(\mathbf{b}^-, \mathbf{a}^+) & \mathbf{\Omega}(\mathbf{c}^-, \mathbf{a}^+) \\ \mathbf{\Omega}(\mathbf{a}^-, \mathbf{b}^+) & \mathbf{\Omega}(\mathbf{b}^-, \mathbf{b}^+) & \mathbf{\Omega}(\mathbf{c}^-, \mathbf{b}^+) \\ \mathbf{\Omega}(\mathbf{a}^-, \mathbf{c}^+) & \mathbf{\Omega}(\mathbf{b}^-, \mathbf{c}^+) & \mathbf{\Omega}(\mathbf{c}^-, \mathbf{c}^+) \end{bmatrix},$$

with $\mathbf{a}^\pm, \mathbf{b}^\pm, \mathbf{c}^\pm$ defined in (5.15). It is assumed to be non-degenerate, $\Pi \neq 0$, and has its sign oriented by taking the quantity $C^+C^-K(0, 0, c, \ell)$ to have a fixed sign where the constants C^+, C^- satisfy

$$\begin{aligned} &e^{-(\mu_1+\mu_2+\mu_3+\mu_4)\xi}\,\widehat{Z}_\xi(\xi) \wedge \mathbf{a}^+(\xi) \wedge \mathbf{b}^+(\xi) \wedge \mathbf{c}^+(\xi) \\ &\to C^+\zeta_1 \wedge \zeta_2 \wedge \zeta_3 \wedge \zeta_4 \text{ as } \xi \to +\infty\,, \end{aligned}$$

$$e^{-(\mu_5+\mu_6+\mu_7+\mu_8)\xi}\,\widehat{Z}_\xi(\xi)\wedge \mathbf{a}^-(\xi)\wedge \mathbf{b}^-(\xi)\wedge \mathbf{c}^-(\xi)$$
$$\to C^-\zeta_5\wedge\zeta_6\wedge\zeta_7\wedge\zeta_8 \text{ as } \xi\to -\infty\,,$$

and the function $K(0,0,c,\ell)$ is defined by

$$\zeta_1(0,0,c,\ell)\wedge\ldots\wedge\zeta_8(0,0,c,\ell) = K(0,0,c,\ell)\varpi\,,\quad \text{with}\quad K\neq 0\,,\quad \forall\,(c,\ell)\in\mathscr{C}\,.$$

(**H6**)—The (λ,ν)-dependent eigenvalues at infinity, determined by

$$\Delta(\mu,\lambda,\nu;c,\ell) := \det\left[\mathbf{B}^\infty(c,\ell)-\lambda\mathbf{M}-\nu\mathbf{N}-\mu(\lambda,\nu,c,\ell)\mathbf{J}(c,\ell)\right]=0\,,$$

are simple and satisfy

$$\mathrm{Re}(\mu_1)\le\mathrm{Re}(\mu_2)\le\mathrm{Re}(\mu_3)\le\mathrm{Re}(\mu_4)<0<\mathrm{Re}(\mu_5)\le\mathrm{Re}(\mu_6)$$
$$\le\mathrm{Re}(\mu_7)\le\mathrm{Re}(\mu_8)\quad\forall\,(\lambda,\nu)\in\Lambda\,,$$

for some open set $\Lambda\subset\mathbb{C}\times\mathbb{R}$ that includes the origin. The set Λ is defined based on the position of the eigenvalues of $\mathbf{A}^\infty(\lambda,\nu)$. The following integral condition is assumed,

$$\int_{-\infty}^{+\infty}\|\mathbf{A}(\xi,\lambda,\nu)-\mathbf{A}^\infty(\lambda,\nu)\|\,\mathrm{d}\xi<+\infty\,,\qquad(\lambda,\nu)\in\Lambda\,,$$

where $\mathbf{A}^\infty(\lambda,\nu)$ is defined in (6.6). We do not assume that the continuous spectrum is stable. However, when the continuous spectrum passes through the origin, we assume that the Evans function is twice differentiable there.

(**H7**)—The Evans function takes the form

$$D(\lambda,\nu):=\frac{1}{2}\det\begin{bmatrix}\mathbf{\Omega}(\mathbf{w}_5,\mathbf{u}_5)&\mathbf{\Omega}(\mathbf{w}_5,\mathbf{u}_6)&\mathbf{\Omega}(\mathbf{w}_5,\mathbf{u}_7)&\mathbf{\Omega}(\mathbf{w}_5,\mathbf{u}_8)\\ \mathbf{\Omega}(\mathbf{w}_6,\mathbf{u}_5)&\mathbf{\Omega}(\mathbf{w}_6,\mathbf{u}_6)&\mathbf{\Omega}(\mathbf{w}_6,\mathbf{u}_7)&\mathbf{\Omega}(\mathbf{w}_6,\mathbf{u}_8)\\ \mathbf{\Omega}(\mathbf{w}_7,\mathbf{u}_5)&\mathbf{\Omega}(\mathbf{w}_7,\mathbf{u}_6)&\mathbf{\Omega}(\mathbf{w}_7,\mathbf{u}_7)&\mathbf{\Omega}(\mathbf{w}_7,\mathbf{u}_8)\\ \mathbf{\Omega}(\mathbf{w}_8,\mathbf{u}_5)&\mathbf{\Omega}(\mathbf{w}_8,\mathbf{u}_6)&\mathbf{\Omega}(\mathbf{w}_8,\mathbf{u}_7)&\mathbf{\Omega}(\mathbf{w}_8,\mathbf{u}_8)\end{bmatrix},$$

for functions $\mathbf{u}_j$, defined in (6.3), and $\mathbf{w}_j$, defined in (6.17), for $j=5,\ldots,8$, and they satisfy the following limits at infinity,

$$\lim_{\xi\to+\infty}e^{-\mu_1(\lambda,\nu)\xi}\mathbf{u}_1(\xi,\lambda,\nu)=\zeta_1(\lambda,\nu)\,,\quad \lim_{\xi\to+\infty}e^{-\mu_2(\lambda,\nu)\xi}\mathbf{u}_2(\xi,\lambda,\nu)=\zeta_2(\lambda,\nu)\,,$$

$$\lim_{\xi\to+\infty}e^{-\mu_3(\lambda,\nu)\xi}\mathbf{u}_3(\xi,\lambda,\nu)=\zeta_3(\lambda,\nu)\,,\quad \lim_{\xi\to+\infty}e^{-\mu_4(\lambda,\nu)\xi}\mathbf{u}_4(\xi,\lambda,\nu)=\zeta_4(\lambda,\nu)\,,$$

$$\lim_{\xi\to-\infty}e^{-\mu_5(\lambda,\nu)\xi}\mathbf{u}_5(\xi,\lambda,\nu)=\zeta_5(\lambda,\nu)\,,\quad \lim_{\xi\to-\infty}e^{-\mu_6(\lambda,\nu)\xi}\mathbf{u}_6(\xi,\lambda,\nu)=\zeta_6(\lambda,\nu)\,,$$

$$\lim_{\xi\to-\infty} e^{-\mu_7(\lambda,\nu)\xi}\mathbf{u}_7(\xi,\lambda,\nu)=\zeta_7(\lambda,\nu)\,,\quad \lim_{\xi\to-\infty} e^{-\mu_8(\lambda,\nu)\xi}\mathbf{u}_8(\xi,\lambda,\nu)=\zeta_8(\lambda,\nu)\,,$$

$$\lim_{\xi\to-\infty} e^{+\mu_1(\lambda,\nu)\xi}\mathbf{w}_1(\xi,\lambda,\nu)=\eta_1(\lambda,\nu)\,,\quad \lim_{\xi\to-\infty} e^{+\mu_2(\lambda,\nu)\xi}\mathbf{w}_2(\xi,\lambda,\nu)=\eta_2(\lambda,\nu)\,,$$

$$\lim_{\xi\to-\infty} e^{+\mu_3(\lambda,\nu)\xi}\mathbf{w}_3(\xi,\lambda,\nu)=\eta_3(\lambda,\nu)\,,\quad \lim_{\xi\to-\infty} e^{+\mu_4(\lambda,\nu)\xi}\mathbf{w}_4(\xi,\lambda,\nu)=\eta_4(\lambda,\nu)\,,$$

$$\lim_{\xi\to+\infty} e^{+\mu_5(\lambda,\nu)\xi}\mathbf{w}_5(\xi,\lambda,\nu)=\eta_5(\lambda,\nu)\,,\quad \lim_{\xi\to+\infty} e^{+\mu_6(\lambda,\nu)\xi}\mathbf{w}_6(\xi,\lambda,\nu)=\eta_6(\lambda,\nu)\,,$$

$$\lim_{\xi\to+\infty} e^{+\mu_7(\lambda,\nu)\xi}\mathbf{w}_7(\xi,\lambda,\nu)=\eta_7(\lambda,\nu)\,,\quad \lim_{\xi\to+\infty} e^{+\mu_8(\lambda,\nu)\xi}\mathbf{w}_8(\xi,\lambda,\nu)=\eta_8(\lambda,\nu)\,,$$

and the following limits when $(\lambda,\nu)\to(0,0)$,

$$\lim_{\lambda,\nu\to0}\mathbf{u}_5(\xi,\lambda,\nu)=\frac{1}{\chi^-}\widehat{Z}_\xi\,,\qquad \lim_{\lambda,\nu\to0}\mathbf{u}_6(\xi,\lambda,\nu)=\mathbf{a}^-(\xi)\,,$$

$$\lim_{\lambda,\nu\to0}\mathbf{u}_7(\xi,\lambda,\nu)=\mathbf{b}^-(\xi)\,,\qquad \lim_{\lambda,\nu\to0}\mathbf{u}_8(\xi,\lambda,\nu)=\mathbf{c}^-(\xi)\,,$$

$$\lim_{\lambda,\nu\to0}\mathbf{w}_5(\xi,\lambda,\nu)=\frac{1}{\chi^+}\widehat{Z}_\xi\,,\qquad \lim_{\lambda,\nu\to0}\mathbf{w}_6(\xi,\lambda,\nu)=\mathbf{a}^+(\xi)\,,$$

$$\lim_{\lambda,\nu\to0}\mathbf{w}_7(\xi,\lambda,\nu)=\mathbf{b}^+(\xi)\,,\qquad \lim_{\lambda,\nu\to0}\mathbf{w}_8(\xi,\lambda,\nu)=\mathbf{c}^+(\xi)\,.$$

Chapter 9
Example: Nonlinear Wave Equation in $2+1$

A class of nonlinear wave equations in $2+1$, of the form (3.49), which is sufficiently general, and also capable of analysis is the following vector-valued wave equation,

$$\Box\phi + \nabla_\phi V = 0 \quad \Box\mathbf{v} + \nabla_{\mathbf{v}} V = 0\,, \tag{9.1}$$

where $\Box$ is the d'Alembertian

$$\Box := \frac{\partial^2}{\partial t^2} - \frac{\partial^2}{\partial x^2} - \frac{\partial^2}{\partial y^2}\,,$$

and $V(\phi, \mathbf{v})$ is a given superquadratic potential. The following potential was constructed by Burchell [34],

$$\begin{aligned} V(\phi, v_1, v_2, v_3) = {} & 2(\phi^2 + {v_1}^2 - {v_2}^2 - {v_3}^2) - 2\phi^3 - \frac{2}{3}{v_1}^3 + {v_2}^3 + \frac{4}{3}{v_3}^3 \\ & -16p(\phi v_1 + v_2 v_3) + \frac{52p}{3}(\phi v_2 + v_1 v_3) + 14p\phi v_3 + 26p v_1 v_2 \\ & -\frac{p}{6}(23\phi^2 + 62{v_1}^2 + 107{v_2}^2 + 68{v_3}^2)\,. \end{aligned} \tag{9.2}$$

This potential has been constructed specifically to yield a class of exact solutions. The derivation of this potential is given in Appendix C. Written out, this system of nonlinear wave equations is

$$\phi_{tt} - \phi_{xx} - \phi_{yy} + 4\phi - 6\phi^2 - p\left(\frac{23}{3}\phi + 16v_1 - \frac{52}{3}v_2 - 14v_3\right) = 0, \tag{9.3}$$

T. J. Burchell, T. J. Bridges, *Transverse Instability of Solitary Waves*, Mathematical Physics Studies, https://doi.org/10.1007/978-981-95-2046-6_9

$$(v_1)_{tt} - (v_1)_{xx} - (v_1)_{yy} + 4v_1 - 2{v_1}^2 - p\left(16\phi + \frac{62}{3}v_1 - 26v_2 - \frac{52}{3}v_3\right) = 0, \tag{9.4}$$

$$(v_2)_{tt} - (v_2)_{xx} - (v_2)_{yy} + 4v_2 - 3{v_2}^2 - p\left(\frac{52}{3}\phi + 26v_1 - \frac{107}{3}v_2 - 16v_3\right) = 0, \tag{9.5}$$

$$(v_3)_{tt} - (v_3)_{xx} - (v_3)_{yy} + 4v_3 - 4{v_3}^2 - p\left(14\phi + \frac{52}{3}v_1 - 16v_2 - \frac{68}{3}v_3\right) = 0, \tag{9.6}$$

with parameter $p > 0$.

Hypothesis (**H1**) is satisfied by rewriting the equations in the form

$$\mathbf{M}Z_t + \mathbf{K}Z_x + \mathbf{N}Z_y = \nabla S(Z), \quad Z \in \mathbb{R}^8 \tag{9.7}$$

with $Z = \left(\phi \ u_1 \ u_2 \ u_3 \ v_1 \ v_2 \ v_3 \ w\right)^T$,

$$\mathbf{M} = \begin{pmatrix} 0 & -1 & 0 & 0 & 0 & 0 & 0 & 0 \\ 1 & 0 & 0 & 0 & 0 & 0 & 0 & 0 \\ 0 & 0 & 0 & 0 & 0 & 0 & 1 & 0 \\ 0 & 0 & 0 & 0 & 0 & -1 & 0 & 0 \\ 0 & 0 & 0 & 0 & 0 & 0 & 0 & -1 \\ 0 & 0 & 0 & 1 & 0 & 0 & 0 & 0 \\ 0 & 0 & -1 & 0 & 0 & 0 & 0 & 0 \\ 0 & 0 & 0 & 0 & 1 & 0 & 0 & 0 \end{pmatrix},$$

$$\mathbf{K} = \begin{pmatrix} 0 & 0 & 1 & 0 & 0 & 0 & 0 & 0 \\ 0 & 0 & 0 & 0 & 0 & 0 & -1 & 0 \\ -1 & 0 & 0 & 0 & 0 & 0 & 0 & 0 \\ 0 & 0 & 0 & 0 & -1 & 0 & 0 & 0 \\ 0 & 0 & 0 & 1 & 0 & 0 & 0 & 0 \\ 0 & 0 & 0 & 0 & 0 & 0 & 0 & -1 \\ 0 & 1 & 0 & 0 & 0 & 0 & 0 & 0 \\ 0 & 0 & 0 & 0 & 0 & 1 & 0 & 0 \end{pmatrix},$$

$$\mathbf{N} = \begin{pmatrix} 0 & 0 & 0 & 1 & 0 & 0 & 0 & 0 \\ 0 & 0 & 0 & 0 & 0 & 1 & 0 & 0 \\ 0 & 0 & 0 & 0 & 1 & 0 & 0 & 0 \\ -1 & 0 & 0 & 0 & 0 & 0 & 0 & 0 \\ 0 & 0 & -1 & 0 & 0 & 0 & 0 & 0 \\ 0 & -1 & 0 & 0 & 0 & 0 & 0 & 0 \\ 0 & 0 & 0 & 0 & 0 & 0 & 0 & -1 \\ 0 & 0 & 0 & 0 & 0 & 0 & 1 & 0 \end{pmatrix},$$

where

$$u_1 = \phi_t + (v_2)_y - (v_3)_x, \qquad u_2 = \phi_x - (v_1)_y - (v_3)_t,$$
$$u_3 = \phi_y + (v_1)_x + (v_2)_t, \qquad w = (v_1)_t + (v_2)_x + (v_3)_y$$

and

$$S(Z) = \frac{1}{2}({u_1}^2 - {u_2}^2 - {u_3}^2 + w^2) + V(\phi, v_1, v_2, v_3). \tag{9.8}$$

The symplectic operator $\mathbf{J}(c, \ell)$ is

$$\mathbf{J} = \mathbf{K} + c\mathbf{M} + \ell\mathbf{N} = \begin{pmatrix} 0 & -c & 1 & \ell & 0 & 0 & 0 & 0 \\ c & 0 & 0 & 0 & 0 & \ell & -1 & 0 \\ -1 & 0 & 0 & 0 & \ell & 0 & c & 0 \\ -\ell & 0 & 0 & 0 & -1 & -c & 0 & 0 \\ 0 & 0 & -\ell & 1 & 0 & 0 & 0 & -c \\ 0 & -\ell & 0 & c & 0 & 0 & 0 & -1 \\ 0 & 1 & -c & 0 & 0 & 0 & 0 & -\ell \\ 0 & 0 & 0 & 0 & c & 1 & \ell & 0 \end{pmatrix},$$

with $\det(\mathbf{J}(c, \ell)) = (1 - c^2 + \ell^2)^4$. The other part of hypothesis (**H1**) will be satisfied provided that $c \neq \pm\sqrt{1 + \ell^2}$. This condition will be met by the definition of the set $\mathscr{C}$ given below for the existence of solitary waves. The resulting solitary wave equation

$$\mathbf{J}(c, \ell)\widehat{Z}_\xi = \nabla S(\widehat{Z}) \tag{9.9}$$

is reversible as the involution

$$\mathbf{R} = \mathrm{diag}(1, -1, -1, -1, 1, 1, 1, -1) \tag{9.10}$$

satisfies (4.15). An exact solitary wave solution is given by

$$\phi = \hat{\phi}, \quad v_1 = 3\hat{\phi}, \quad v_2 = 2\hat{\phi}, \quad v_3 = \frac{3}{2}\hat{\phi}, \tag{9.11}$$

where

$$\hat{\phi} := \operatorname{sech}^2(\alpha\xi), \quad \xi = x + \ell y + ct, \quad \alpha = \frac{1}{\sqrt{1+\ell^2-c^2}}. \tag{9.12}$$

Hence the set $\mathscr{C}$ is taken to be

$$\mathscr{C} = \left\{(c,\ell) \in \mathbb{R}^2 : -\sqrt{1+\ell^2} < c < \sqrt{1+\ell^2}\right\}, \tag{9.13}$$

which is the space between the branches of a hyperbola in (c, ℓ)-space.

In terms of the Z-coordinates the solitary wave solution is

$$\widehat{Z}(\xi, c, \ell) = \begin{pmatrix} \hat{\phi} \\ (c + 2\ell - \frac{3}{2})\hat{\phi}_\xi \\ (1 - \frac{3}{2}c - 3\ell)\hat{\phi}_\xi \\ (3 + 2c + \ell)\hat{\phi}_\xi \\ 3\hat{\phi} \\ 2\hat{\phi} \\ \frac{3}{2}\hat{\phi} \\ (2 + 3c + \frac{3}{2}\ell)\hat{\phi}_\xi \end{pmatrix}. \tag{9.14}$$

This construction satisfies the first part of hypothesis (**H2**), while it is also clear that the solitary wave is a smooth function of c, ℓ and ξ for $(c, \ell) \in \mathscr{C}$. Non-degeneracy of the momentum derivatives will be verified in Sect. 9.3. This solitary wave is also reversible since it satisfies (4.16).

9.1 Linearisation About the Solitary Wave

The linearisation about the solitary wave solution is

$$\mathbf{J}Z_\xi = [\mathbf{B}(\xi, c, \ell) - \lambda\mathbf{M} - \mathrm{i}\nu\mathbf{N}]\, Z, \quad Z \in \mathbb{C}^8, \tag{9.15}$$

where

$$\mathbf{B}(\xi, c, \ell) = \begin{pmatrix} -\widehat{\varphi} - \frac{23}{3}p & 0 & 0 & 0 & -16p & \frac{52}{3}p & 14p & 0 \\ 0 & 1 & 0 & 0 & 0 & 0 & 0 & 0 \\ 0 & 0 & -1 & 0 & 0 & 0 & 0 & 0 \\ 0 & 0 & 0 & -1 & 0 & 0 & 0 & 0 \\ -16p & 0 & 0 & 0 & -\widehat{\varphi} - \frac{62}{3}p & 26p & \frac{52}{3}p & 0 \\ \frac{52}{3}p & 0 & 0 & 0 & 26p & \widehat{\varphi} - \frac{107}{3}p & -16p & 0 \\ 14p & 0 & 0 & 0 & \frac{52}{3}p & -16p & \widehat{\varphi} - \frac{68}{3}p & 0 \\ 0 & 0 & 0 & 0 & 0 & 0 & 0 & 1 \end{pmatrix}, \tag{9.16}$$

where $\widehat{\varphi} = 12\hat{\phi} - 4$. The system at infinity is

$$\mathbf{J}Z_\xi = \left[\mathbf{B}^\infty(c, \ell) - \lambda\mathbf{M} - \mathrm{i}\nu\mathbf{N}\right] Z, \quad Z \in \mathbb{C}^8, \tag{9.17}$$

with $\mathbf{B}^\infty(c, \ell)$ the same as (9.16) but with $\widehat{\varphi}$ replaced with -4. The eigenvalues of the system at infinity are defined by $\Delta(\mu, \lambda, \nu) = 0$, with

$$\begin{aligned} \Delta(\mu, \lambda, \nu) &= \det[\mathbf{B}^\infty(c, \ell) - \lambda\mathbf{M} - \mathrm{i}\nu\mathbf{N} - \mu\mathbf{J}(c, \ell)] \\ &= \Theta^4 - (16 + 30p)\Theta^3 + (96 + 360p + 275p^2)\Theta^2 \\ &\quad - (256 + 1440p + 2200p^2 + 750p^3)\Theta \\ &\quad + 256 + 1920p + 4400p^2 + 3000p^3 \\ &= (\Theta - 4)(\Theta - 4 - 5p)(\Theta - 4 - 10p)(\Theta - 4 - 15p)\,, \end{aligned}$$

where

$$\Theta(\mu, \lambda, \nu) = \mu^2 - (c\mu + \lambda)^2 + (\ell\mu + \mathrm{i}\nu)^2.$$

From this we can determine the continuous spectrum as

$$\sigma^{\text{cont}} = \{\lambda \in \mathbb{C}, \nu \in \mathbb{R} : \Delta(\mathrm{i}\kappa, \lambda, \nu) = 0, \kappa \in \mathbb{R}\}, \tag{9.18}$$

which can be separated into 4 parts by considering the 4 solutions of the equation for Θ:

$$\Theta(\mathrm{i}\kappa, \lambda, \nu) = 4, 4 + 5p, 4 + 10p, 4 + 15p$$

to get

$$\sigma^{\text{cont}} = \bigcup_{n=0}^{3} \{\lambda \in \mathbb{C}, \nu \in \mathbb{R} : \Theta(\mathrm{i}\kappa, \lambda, \nu) = 4 + 5np, \kappa \in \mathbb{R}\}. \tag{9.19}$$

Now if we solve the equation $\Theta(\mathrm{i}\kappa, \lambda, \nu) = 4 + 5np$ for λ in terms of κ, ν we find that

$$\lambda = -\mathrm{i}c\kappa \pm \mathrm{i}\sqrt{\kappa^2 + (\ell\kappa + \nu)^2 + 4 + 5np}\,,$$

which is purely imaginary for all $\kappa, \nu \in \mathbb{R}$ since $np \geq 0$, $n = 0, \ldots, 3$. This means that the whole of the continuous spectrum will be contained within the plane $(Im(\lambda), \nu)$.

Let $\tilde{\lambda} = Im(\lambda)$. Then the equation $\Theta(\mathrm{i}\kappa, \mathrm{i}\tilde{\lambda}, \nu) = 4 + 5np$ can be rearranged as a quadratic equation for κ with real coefficients:

$$\begin{aligned} &\kappa^2 - (c\kappa + \tilde{\lambda})^2 + (\ell\kappa + \nu)^2 + 4 + 5np = 0, \\ \Rightarrow \quad &\alpha^{-2}\kappa^2 + 2(\ell\nu - c\tilde{\lambda})\kappa - \tilde{\lambda}^2 + \nu^2 + 4 + 5np = 0. \end{aligned}$$

Since this equation is required to have real solutions we can deduce that

$$\begin{aligned} &(\ell\nu - c\tilde{\lambda})^2 \geq \alpha^{-2}(-\tilde{\lambda}^2 + \nu^2 + 4 + 5np) \\ \Rightarrow \quad &(1 + \ell^2)\tilde{\lambda}^2 - 2c\ell\tilde{\lambda}\nu + (c^2 - 1)\nu^2 \geq \alpha^{-2}(4 + 5np). \end{aligned}$$

Hence the continuous spectrum can also be written as

$$\begin{aligned} \sigma^{\text{cont}} = \bigcup_{n=0}^{3} \{\lambda = \mathrm{i}\tilde{\lambda} \in \mathrm{i}\mathbb{R}, \nu \in \mathbb{R} : (1 + \ell^2)\tilde{\lambda}^2 - 2c\ell\tilde{\lambda}\nu + (c^2 - 1)\nu^2 \\ \geq \alpha^{-2}(4 + 5np)\}. \end{aligned} \tag{9.20}$$

9.2 Eigenvectors and the Factor χ

Now solving the equation $\Delta(\mu, 0, 0) = 0$ allows us to find the 8 μ-eigenvalues

$$\mu_1 = -\sqrt{\frac{4 + 15p}{1 - c^2 + \ell^2}}, \qquad \mu_3 = -\sqrt{\frac{4 + 5p}{1 - c^2 + \ell^2}},$$

$$\mu_5 = \sqrt{\frac{4}{1 - c^2 + \ell^2}}, \qquad \mu_7 = \sqrt{\frac{4 + 10p}{1 - c^2 + \ell^2}},$$

$$\mu_2 = -\sqrt{\frac{4 + 10p}{1 - c^2 + \ell^2}}, \qquad \mu_4 = -\sqrt{\frac{4}{1 - c^2 + \ell^2}},$$

$$\mu_6 = \sqrt{\frac{4 + 5p}{1 - c^2 + \ell^2}}, \qquad \mu_8 = \sqrt{\frac{4 + 15p}{1 - c^2 + \ell^2}}.$$

With $(c, \ell) \in \mathscr{C}$ they are simple and real, while the tangent vector $\widehat{Z}_\xi$ is bi-asymptotic to the eigenvalues μ_4 and μ_5. The eigenvectors and adjoint eigenvectors for $\lambda, \mu = 0$ are

$$\zeta_1 = z_1 \begin{pmatrix} \frac{3}{2} \\ \mu_1(\frac{3}{2}c + 3\ell - 1) \\ \mu_1(\frac{3}{2} - c - 2\ell) \\ \mu_1(2 + 3c + \frac{3}{2}\ell) \\ 2 \\ 3 \\ 1 \\ \mu_1(3 + 2c + \ell) \end{pmatrix}, \qquad \eta_1 = h_1 \begin{pmatrix} \frac{3}{2} \\ \mu_1(1 - \frac{3}{2}c - 3\ell) \\ \mu_1(c + 2\ell - \frac{3}{2}) \\ -\mu_1(2 + 3c + \frac{3}{2}\ell) \\ 2 \\ 3 \\ 1 \\ -\mu_1(3 + 2c + \ell) \end{pmatrix},$$

$$\zeta_2 = z_2 \begin{pmatrix} 2 \\ \mu_2(2c + \ell - 3) \\ \mu_2(2 - 3c - \frac{3}{2}\ell) \\ \mu_2(\frac{3}{2} + c + 2\ell) \\ \frac{3}{2} \\ 1 \\ 3 \\ \mu_2(1 + \frac{3}{2}c + 3\ell) \end{pmatrix}, \qquad \eta_2 = h_2 \begin{pmatrix} 2 \\ \mu_2(3 - 2c - \ell) \\ \mu_2(3c + \frac{3}{2}\ell - 2) \\ -\mu_2(\frac{3}{2} + c + 2\ell) \\ \frac{3}{2} \\ 1 \\ 3 \\ -\mu_2(1 + \frac{3}{2}c + 3\ell) \end{pmatrix},$$

$$\zeta_3 = z_3 \begin{pmatrix} 3 \\ \mu_3(3c + \frac{3}{2}\ell - 2) \\ \mu_3(3 - 2c - \ell) \\ \mu_3(1 + \frac{3}{2}c + 3\ell) \\ 1 \\ \frac{3}{2} \\ 2 \\ \mu_3(\frac{3}{2} + c + 2\ell) \end{pmatrix}, \qquad \eta_3 = h_3 \begin{pmatrix} 3 \\ \mu_3(2 - 3c - \frac{3}{2}\ell) \\ \mu_3(2c + \ell - 3) \\ -\mu_3(1 + \frac{3}{2}c + 3\ell) \\ 1 \\ \frac{3}{2} \\ 2 \\ -\mu_3(\frac{3}{2} + c + 2\ell) \end{pmatrix},$$

$$\zeta_4 = z_4 \begin{pmatrix} 1 \\ \mu_4(c + 2\ell - \frac{3}{2}) \\ \mu_4(1 - \frac{3}{2}c - 3\ell) \\ \mu_4(3 + 2c + \ell) \\ 3 \\ 2 \\ \frac{3}{2} \\ \mu_4(2 + 3c + \frac{3}{2}\ell) \end{pmatrix}, \qquad \eta_4 = h_4 \begin{pmatrix} 1 \\ \mu_4(\frac{3}{2} - c - 2\ell) \\ \mu_4(\frac{3}{2}c + 3\ell - 1) \\ -\mu_4(3 + 2c + \ell) \\ 3 \\ 2 \\ \frac{3}{2} \\ -\mu_4(2 + 3c + \frac{3}{2}\ell) \end{pmatrix},$$

$$\zeta_5 = z_5 \begin{pmatrix} 1 \\ \mu_5(c + 2\ell - \frac{3}{2}) \\ \mu_5(1 - \frac{3}{2}c - 3\ell) \\ \mu_5(3 + 2c + \ell) \\ 3 \\ 2 \\ \frac{3}{2} \\ \mu_5(2 + 3c + \frac{3}{2}\ell) \end{pmatrix}, \qquad \eta_5 = h_5 \begin{pmatrix} 1 \\ \mu_5(\frac{3}{2} - c - 2\ell) \\ \mu_5(\frac{3}{2}c + 3\ell - 1) \\ -\mu_5(3 + 2c + \ell) \\ 3 \\ 2 \\ \frac{3}{2} \\ -\mu_5(2 + 3c + \frac{3}{2}\ell) \end{pmatrix},$$

$$\zeta_6 = z_6 \begin{pmatrix} 3 \\ \mu_6(3c + \frac{3}{2}\ell - 2) \\ \mu_6(3 - 2c - \ell) \\ \mu_6(1 + \frac{3}{2}c + 3\ell) \\ 1 \\ \frac{3}{2} \\ 2 \\ \mu_6(\frac{3}{2} + c + 2\ell) \end{pmatrix}, \qquad \eta_6 = h_6 \begin{pmatrix} 3 \\ \mu_6(2 - 3c - \frac{3}{2}\ell) \\ \mu_6(2c + \ell - 3) \\ -\mu_6(1 + \frac{3}{2}c + 3\ell) \\ 1 \\ \frac{3}{2} \\ 2 \\ -\mu_6(\frac{3}{2} + c + 2\ell) \end{pmatrix},$$

$$\zeta_7 = z_7 \begin{pmatrix} 2 \\ \mu_7(2c + \ell - 3) \\ \mu_7(2 - 3c - \frac{3}{2}\ell) \\ \mu_7(\frac{3}{2} + c + 2\ell) \\ \frac{3}{2} \\ 1 \\ 3 \\ \mu_7(1 + \frac{3}{2}c + 3\ell) \end{pmatrix}, \qquad \eta_7 = h_7 \begin{pmatrix} 2 \\ \mu_7(3 - 2c - \ell) \\ \mu_7(3c + \frac{3}{2}\ell - 2) \\ -\mu_7(\frac{3}{2} + c + 2\ell) \\ \frac{3}{2} \\ 1 \\ 3 \\ -\mu_7(1 + \frac{3}{2}c + 3\ell) \end{pmatrix},$$

$$\zeta_8 = z_8 \begin{pmatrix} \frac{3}{2} \\ \mu_8(\frac{3}{2}c + 3\ell - 1) \\ \mu_8(\frac{3}{2} - c - 2\ell) \\ \mu_8(2 + 3c + \frac{3}{2}\ell) \\ 2 \\ 3 \\ 1 \\ \mu_8(3 + 2c + \ell) \end{pmatrix}, \qquad \eta_8 = h_8 \begin{pmatrix} \frac{3}{2} \\ \mu_8(1 - \frac{3}{2}c - 3\ell) \\ \mu_8(c + 2\ell - \frac{3}{2}) \\ -\mu_8(2 + 3c + \frac{3}{2}\ell) \\ 2 \\ 3 \\ 1 \\ -\mu_8(3 + 2c + \ell) \end{pmatrix},$$

with z_j, h_j arbitrary real numbers. The normalisation $\mathbf{\Omega}(\eta_i, \zeta_j) = \delta_{ij}$ then yields the constraints

$$h_1 z_1 = \frac{-2\alpha}{15\sqrt{4 + 15p}}, \qquad h_2 z_2 = \frac{-2\alpha}{15\sqrt{4 + 10p}},$$

$$h_3 z_3 = \frac{2\alpha}{15\sqrt{4 + 5p}}, \qquad h_4 z_4 = \frac{\alpha}{15},$$

$$h_5 z_5 = \frac{-\alpha}{15}, \qquad h_6 z_6 = \frac{-2\alpha}{15\sqrt{4+5p}},$$
$$h_7 z_7 = \frac{2\alpha}{15\sqrt{4+10p}}, \qquad h_8 z_8 = \frac{2\alpha}{15\sqrt{4+15p}}. \tag{9.21}$$

Since the z's and h's are arbitrary, take

$$\begin{array}{llll} z_1 = \sigma_1\,, & z_2 = \sigma_2\,, & z_3 = \sigma_3\,, & z_4 = \sigma_4\,, \\ z_5 = -\sigma_4\,, & z_6 = -\sigma_3\,, & z_7 = \sigma_2\,, & z_8 = \sigma_1\,, \\ h_1 = -\sigma_1\,, & h_2 = -\sigma_2\,, & h_3 = \sigma_3\,, & h_4 = \sigma_4\,, \\ h_5 = \sigma_4\,, & h_6 = \sigma_3\,, & h_7 = \sigma_2\,, & h_8 = \sigma_1\,, \end{array} \tag{9.22}$$

where

$$\sigma_1 = \sqrt{\frac{2\alpha}{15\sqrt{4+15p}}}, \quad \sigma_2 = \sqrt{\frac{2\alpha}{15\sqrt{4+10p}}}, \quad \sigma_3 = \sqrt{\frac{2\alpha}{15\sqrt{4+5p}}}, \quad \sigma_4 = \sqrt{\frac{\alpha}{15}}.$$

This confirms hypothesis (**H4**). Since the solitary wave is reversible and

$$\mathbf{\Omega}(\zeta_5, \mathbf{R}\zeta_5) = \frac{15}{2}\mu_5(\sigma_4)^2(1+\ell^2-c^2) = 1\,, \tag{9.23}$$

it follows from Proposition 5.3 that $\chi > 0$. An explicit formula for χ can be found by checking the asymptotics of the solitary wave

$$\lim_{\xi\to-\infty} e^{-\mu_5\xi}\widehat{Z}_\xi = -\frac{8\alpha}{\sigma_4}\zeta_5 \quad \text{and} \quad \lim_{\xi\to+\infty} e^{+\mu_5\xi}\widehat{Z}_\xi = -\frac{8\alpha}{\sigma_4}\eta_5\,.$$

Hence

$$\chi^-\chi^+ = \left(-\frac{8\alpha}{\sigma_4}\right)\left(-\frac{8\alpha}{\sigma_4}\right) = \frac{64\alpha^2}{(\sigma_4)^2} = 960\alpha \quad \Rightarrow \quad \chi = \frac{1}{960\alpha}\,. \tag{9.24}$$

9.3 Momentum Derivatives

To compute the momentum derivatives, first note that I_1, I_2 in this case are

$$\begin{aligned} I_1 &= \frac{1}{2}\int_{-\infty}^{+\infty} \langle \mathbf{M}\widehat{Z}_\xi, \widehat{Z}\rangle \,\mathrm{d}\xi \\ &= \frac{1}{2}\int_{-\infty}^{+\infty} -\left(c+2\ell-\frac{3}{2}\right)\hat{\phi}\hat{\phi}_{\xi\xi} + \left(c+2\ell-\frac{3}{2}\right)\hat{\phi}_\xi^2 + \frac{3}{2}\left(1-\frac{3}{2}c-3\ell\right)\hat{\phi}_\xi^2 \end{aligned}$$

$$- 2(3+2c+\ell)\hat{\phi}_\xi^2 - 3\left(2+3c+\frac{3}{2}\ell\right)\hat{\phi}\hat{\phi}_{\xi\xi} + 2(3+2c+\ell)\hat{\phi}\hat{\phi}_{\xi\xi}$$

$$- \frac{3}{2}\left(1-\frac{3}{2}c-3\ell\right)\hat{\phi}\hat{\phi}_{\xi\xi} + 3\left(2+3c+\frac{3}{2}\ell\right)\hat{\phi}_\xi^2 \,\mathrm{d}\xi$$

$$= \frac{15c}{8}\int_{-\infty}^{+\infty}\hat{\phi}_\xi^2 - \hat{\phi}\hat{\phi}_{\xi\xi}\,\mathrm{d}\xi$$

$$= \frac{15c}{4}\int_{-\infty}^{+\infty}\hat{\phi}_\xi^2\,\mathrm{d}\xi$$

$$= \frac{4c}{\sqrt{1+\ell^2-c^2}} \tag{9.25}$$

and

$$I_2 = \frac{1}{2}\int_{-\infty}^{+\infty}\langle \mathbf{N}\widehat{Z}_\xi, \widehat{Z}\rangle \,\mathrm{d}\xi$$

$$= \frac{1}{2}\int_{-\infty}^{+\infty}(3+2c+\ell)\hat{\phi}\hat{\phi}_{\xi\xi} + 2\left(c+2\ell-\frac{3}{2}\right)\hat{\phi}_\xi^2 + 3\left(1-\frac{3}{2}c-3\ell\right)\hat{\phi}_\xi^2$$

$$- (3+2c+\ell)\hat{\phi}_\xi^2 - 3\left(1-\frac{3}{2}c-3\ell\right)\hat{\phi}\hat{\phi}_{\xi\xi} - 2\left(c+2\ell-\frac{3}{2}\right)\hat{\phi}\hat{\phi}_{\xi\xi}$$

$$- \frac{3}{2}\left(2+3c+\frac{3}{2}\ell\right)\hat{\phi}\hat{\phi}_{\xi\xi} + \frac{3}{2}\left(2+3c+\frac{3}{2}\ell\right)\hat{\phi}_\xi^2\,\mathrm{d}\xi$$

$$= \frac{15\ell}{8}\int_{-\infty}^{+\infty}\hat{\phi}\hat{\phi}_{\xi\xi} - \hat{\phi}_\xi^2\,\mathrm{d}\xi$$

$$= -\frac{15\ell}{4}\int_{-\infty}^{+\infty}\hat{\phi}_\xi^2\,\mathrm{d}\xi$$

$$= -\frac{4\ell}{\sqrt{1+\ell^2-c^2}} \tag{9.26}$$

after integrating by parts and using the exact solution (9.12) and its integral

$$\int_{-\infty}^{+\infty}\hat{\phi}_\xi^2\,\mathrm{d}\xi = 4\alpha^2\int_{-\infty}^{+\infty}\tanh^2(\alpha\xi)\mathrm{sech}^4(\alpha\xi)\,\mathrm{d}\xi = \frac{16}{15}(1+\ell^2-c^2)^{-1/2}\,.$$

Differentiating with respect to c and ℓ then gives us

$$\frac{\partial I_1}{\partial c} = \frac{4(1+\ell^2)}{(1+\ell^2-c^2)^{\frac{3}{2}}} = 4(1+\ell^2)\alpha^3, \tag{9.27}$$

$$\frac{\partial I_1}{\partial \ell} = \frac{\partial I_2}{\partial c} = \frac{-4c\ell}{(1+\ell^2-c^2)^{\frac{3}{2}}} = -4c\ell\alpha^3, \tag{9.28}$$

$$\frac{\partial I_2}{\partial \ell} = \frac{4(c^2-1)}{(1+\ell^2-c^2)^{\frac{3}{2}}} = 4(c^2-1)\alpha^3. \tag{9.29}$$

We can now verify that (4.11) holds,

$$\begin{vmatrix} 4(1+\ell^2)\alpha^3 & -4c\ell\alpha^3 \\ -4c\ell\alpha^3 & 4(c^2-1)\alpha^3 \end{vmatrix} = -16\alpha^4 \neq 0, \quad \forall (c,\ell) \in \mathscr{C}, \tag{9.30}$$

thus confirming the rest of hypothesis (**H2**).

9.4 The Lazutkin–Treschev Invariant

In order to work out the Lazutkin–Treschev invariant Π, we need to find expressions for the Jost solutions $\mathbf{a}^{\pm}, \mathbf{b}^{\pm}, \mathbf{c}^{\pm}$. First, express the solutions of (9.15) by $Z = \left(\tilde{\phi}\ \tilde{u}_1\ \tilde{u}_2\ \tilde{u}_3\ \tilde{v}_1\ \tilde{v}_2\ \tilde{v}_3\ \tilde{w}\right)^T$. Then $\tilde{u}_1, \tilde{u}_2, \tilde{u}_3$ and $\tilde{w}$ can be obtained from (9.15) as

$$\begin{aligned}
\tilde{u}_1 &= \lambda\tilde{\phi} + c\tilde{\phi}_\xi + \mathtt{i}\nu\tilde{v}_2 + \ell(\tilde{v}_2)_\xi - (\tilde{v}_3)_\xi,\\
\tilde{u}_2 &= \tilde{\phi}_\xi - \mathtt{i}\nu\tilde{v}_1 - \ell(\tilde{v}_1)_\xi - \lambda\tilde{v}_3 - c(\tilde{v}_3)_\xi,\\
\tilde{u}_3 &= \mathtt{i}\nu\tilde{\phi} + \ell\tilde{\phi}_\xi + (\tilde{v}_1)_\xi + \lambda\tilde{v}_2 + c(\tilde{v}_2)_\xi,\\
\tilde{w} &= \lambda\tilde{v}_1 + c(\tilde{v}_1)_\xi + (\tilde{v}_2)_\xi + \mathtt{i}\nu\tilde{v}_3 + \ell(\tilde{v}_3)_\xi.
\end{aligned}$$

This allows us to reduce (9.15) to a set of 4 coupled equations:

$$L\tilde{\phi} + p\left(\frac{23}{3}\tilde{\phi} + 16\tilde{v}_1 - \frac{52}{3}\tilde{v}_2 - 14\tilde{v}_3\right) = 0, \tag{9.31}$$

$$L\tilde{v}_1 + p\left(16\tilde{\phi} + \frac{62}{3}\tilde{v}_1 - 26\tilde{v}_2 - \frac{52}{3}\tilde{v}_3\right) = 0, \tag{9.32}$$

$$L\tilde{v}_2 + p\left(\frac{52}{3}\tilde{\phi} + 26\tilde{v}_1 - \frac{107}{3}\tilde{v}_2 - 16\tilde{v}_3\right) = 0, \tag{9.33}$$

$$L\tilde{v}_3 + p\left(14\tilde{\phi} + \frac{52}{3}\tilde{v}_1 - 16\tilde{v}_2 - \frac{68}{3}\tilde{v}_3\right) = 0, \tag{9.34}$$

where

$$L = \alpha^{-2}\partial_{\xi\xi} - 2(c\lambda - \mathtt{i}\ell\nu)\partial_\xi + \widehat{\varphi} - \lambda^2 - \nu^2.$$

If we now take the transformation

$$\begin{aligned}\tilde{\phi} &= e^{\alpha^2\Upsilon\xi}\left(\psi_1 + 3\psi_2 + 2\psi_3 + \frac{3}{2}\psi_4\right),\\ \tilde{v}_1 &= e^{\alpha^2\Upsilon\xi}\left(3\psi_1 + \psi_2 + \frac{3}{2}\psi_3 + 2\psi_4\right),\\ \tilde{v}_2 &= e^{\alpha^2\Upsilon\xi}\left(2\psi_1 + \frac{3}{2}\psi_2 + \psi_3 + 3\psi_4\right),\\ \tilde{v}_3 &= e^{\alpha^2\Upsilon\xi}\left(\frac{3}{2}\psi_1 + 2\psi_2 + 3\psi_3 + \psi_4\right),\end{aligned} \tag{9.35}$$

(where $\Upsilon = c\lambda - \mathtt{i}\ell\nu$) then Eqs. (9.31)–(9.34) will decouple to give

$$\alpha^{-2}(\psi_1)_{\xi\xi} + \left(\widehat{\varphi} - \alpha^2\Upsilon^2 - \lambda^2 - \nu^2\right)\psi_1 = 0, \tag{9.36}$$

$$\alpha^{-2}(\psi_2)_{\xi\xi} + \left(\widehat{\varphi} - 5p - \alpha^2\Upsilon^2 - \lambda^2 - \nu^2\right)\psi_2 = 0, \tag{9.37}$$

$$\alpha^{-2}(\psi_3)_{\xi\xi} + \left(\widehat{\varphi} - 10p - \alpha^2\Upsilon^2 - \lambda^2 - \nu^2\right)\psi_3 = 0, \tag{9.38}$$

$$\alpha^{-2}(\psi_4)_{\xi\xi} + \left(\widehat{\varphi} - 15p - \alpha^2\Upsilon^2 - \lambda^2 - \nu^2\right)\psi_4 = 0. \tag{9.39}$$

The derivation of this transformation is given in Appendix C.

In order to construct the Jost solutions $\mathbf{a}^{\pm}, \mathbf{b}^{\pm}, \mathbf{c}^{\pm}$ we only need to consider the solutions of these equations when $\lambda, \nu = 0$, in which case they simplify to

$$\alpha^{-2}(\psi_1)_{\xi\xi} + 12\,\mathrm{sech}^2(\alpha\xi)\psi_1 = 4\psi_1, \tag{9.40}$$

$$\alpha^{-2}(\psi_2)_{\xi\xi} + 12\,\mathrm{sech}^2(\alpha\xi)\psi_2 = (4 + 5p)\psi_2, \tag{9.41}$$

$$\alpha^{-2}(\psi_3)_{\xi\xi} + 12\,\mathrm{sech}^2(\alpha\xi)\psi_3 = (4 + 10p)\psi_3, \tag{9.42}$$

$$\alpha^{-2}(\psi_4)_{\xi\xi} + 12\,\mathrm{sech}^2(\alpha\xi)\psi_4 = (4 + 15p)\psi_4. \tag{9.43}$$

Now each of these equations has the general form

$$\alpha^{-2}\Psi_{\xi\xi} + 12\,\mathrm{sech}^2(\alpha\xi)\Psi = \kappa^2\Psi,$$

which has solutions:

$$\Psi = e^{\mp\alpha\kappa\xi}\left[\pm\frac{\kappa(\kappa^2 - 4)}{15} + \frac{2\kappa^2 - 3}{5}\tanh(\alpha\xi) \pm \kappa\tanh^2(\alpha\xi) + \tanh^3(\alpha\xi)\right].$$

The two solutions of (9.40) are proportional to $\hat{\phi}_\xi$ and

$$15\alpha\xi\tanh(\alpha\xi)\,\mathrm{sech}^2(\alpha\xi) - 8\,\mathrm{sech}^2(\alpha\xi) + 7\tanh^2(\alpha\xi) + 2\sinh^2(\alpha\xi)\,.$$

For the other six solutions, we proceed as follows. By considering the eigenvalues of the system and how they correspond to $\mathbf{a}^{\pm}$, $\mathbf{b}^{\pm}$, $\mathbf{c}^{\pm}$ we can deduce that $\mathbf{a}^{\pm}$ will be generated by the solutions of (9.41)

$$\psi_{\mathbf{a}}^{\pm} = e^{\mp\alpha\kappa_1\xi}\left[\pm\frac{p\kappa_1}{3} + (1+2p)\tanh(\alpha\xi) \pm \kappa_1\tanh^2(\alpha\xi) + \tanh^3(\alpha\xi)\right],$$

$\mathbf{b}^{\pm}$ will be generated by the solutions of (9.42)

$$\psi_{\mathbf{b}}^{\pm} = e^{\mp\alpha\kappa_2\xi}\left[\pm\frac{2p\kappa_2}{3} + (1+4p)\tanh(\alpha\xi) \pm \kappa_2\tanh^2(\alpha\xi) + \tanh^3(\alpha\xi)\right],$$

and $\mathbf{c}^{\pm}$ will be generated by the solutions of (9.43)

$$\psi_{\mathbf{c}}^{\pm} = e^{\mp\alpha\kappa_3\xi}\left[\pm p\kappa_3 + (1+6p)\tanh(\alpha\xi) \pm \kappa_3\tanh^2(\alpha\xi) + \tanh^3(\alpha\xi)\right],$$

where $\kappa_1 = \sqrt{4+5p}$, $\kappa_2 = \sqrt{4+10p}$ and $\kappa_3 = \sqrt{4+15p}$.

The only one of the eight solutions to be in $L^2(\mathbb{R})$ is $\hat{\phi}_\xi$ (which will generate $\widehat{Z}_\xi$); hence the kernel condition (5.6) can be verified, confirming hypothesis (**H3**).

By reversing the transformations to obtain the Z-coordinates we find that

$$\mathbf{a}^{\pm} = \mathcal{P}_{\mathbf{a}}\begin{pmatrix} 3\psi_{\mathbf{a}}^{\pm} \\ (3c+\frac{3}{2}\ell-2)(\psi_{\mathbf{a}}^{\pm})_\xi \\ (3-2c-\ell)(\psi_{\mathbf{a}}^{\pm})_\xi \\ (1+\frac{3}{2}c+3\ell)(\psi_{\mathbf{a}}^{\pm})_\xi \\ \psi_{\mathbf{a}}^{\pm} \\ \frac{3}{2}\psi_{\mathbf{a}}^{\pm} \\ 2\psi_{\mathbf{a}}^{\pm} \\ (\frac{3}{2}+c+2\ell)(\psi_{\mathbf{a}}^{\pm})_\xi \end{pmatrix},$$

$$\mathbf{b}^{\pm} = \pm\mathcal{P}_{\mathbf{b}}\begin{pmatrix} 2\psi_{\mathbf{b}}^{\pm} \\ (2c+\ell-3)(\psi_{\mathbf{b}}^{\pm})_\xi \\ (2-3c-\frac{3}{2}\ell)(\psi_{\mathbf{b}}^{\pm})_\xi \\ (\frac{3}{2}+c+2\ell)(\psi_{\mathbf{b}}^{\pm})_\xi \\ \frac{3}{2}\psi_{\mathbf{b}}^{\pm} \\ \psi_{\mathbf{b}}^{\pm} \\ 3\psi_{\mathbf{b}}^{\pm} \\ (1+\frac{3}{2}c+3\ell)(\psi_{\mathbf{b}}^{\pm})_\xi \end{pmatrix},$$

$$\mathbf{c}^{\pm} = \pm \mathcal{P}_{\mathbf{c}} \begin{pmatrix} \frac{3}{2}\psi_{\mathbf{c}}^{\pm} \\ (\frac{3}{2}c + 3\ell - 1)(\psi_{\mathbf{c}}^{\pm})_{\xi} \\ (\frac{3}{2} - c - 2\ell)(\psi_{\mathbf{c}}^{\pm})_{\xi} \\ (2 + 3c + \frac{3}{2}\ell)(\psi_{\mathbf{c}}^{\pm})_{\xi} \\ 2\psi_{\mathbf{c}}^{\pm} \\ 3\psi_{\mathbf{c}}^{\pm} \\ \psi_{\mathbf{c}}^{\pm} \\ (3 + 2c + \ell)(\psi_{\mathbf{c}}^{\pm})_{\xi} \end{pmatrix},$$

where

$$\mathcal{P}_{\mathbf{a}} = \frac{3\sigma_3}{(p+3)\sqrt{4+5p} + 6p + 6},$$

$$\mathcal{P}_{\mathbf{b}} = \frac{3\sigma_2}{(2p+3)\sqrt{4+10p} + 12p + 6},$$

$$\mathcal{P}_{\mathbf{c}} = \frac{\sigma_1}{(p+1)\sqrt{4+15p} + 6p + 2}.$$

These coefficients (which are positive and finite for all $p > 0$) ensure that the asymptotic conditions (5.1) hold. This allows us to evaluate:

$$\boldsymbol{\Omega}(\mathbf{a}^-, \mathbf{a}^+) = \frac{15(\mathcal{P}_{\mathbf{a}})^2}{4\alpha^2}\big((\psi_{\mathbf{a}}^-)_{\xi}\psi_{\mathbf{a}}^+ - \psi_{\mathbf{a}}^-(\psi_{\mathbf{a}}^+)_{\xi}\big), \tag{9.44}$$

$$\boldsymbol{\Omega}(\mathbf{a}^-, \mathbf{b}^+) = 0, \tag{9.45}$$

$$\boldsymbol{\Omega}(\mathbf{a}^-, \mathbf{c}^+) = 0, \tag{9.46}$$

$$\boldsymbol{\Omega}(\mathbf{b}^-, \mathbf{a}^+) = 0, \tag{9.47}$$

$$\boldsymbol{\Omega}(\mathbf{b}^-, \mathbf{b}^+) = \frac{15(\mathcal{P}_{\mathbf{b}})^2}{4\alpha^2}\big((\psi_{\mathbf{b}}^-)_{\xi}\psi_{\mathbf{b}}^+ - \psi_{\mathbf{b}}^-(\psi_{\mathbf{b}}^+)_{\xi}\big), \tag{9.48}$$

$$\boldsymbol{\Omega}(\mathbf{b}^-, \mathbf{c}^+) = 0, \tag{9.49}$$

$$\boldsymbol{\Omega}(\mathbf{c}^-, \mathbf{a}^+) = 0, \tag{9.50}$$

$$\boldsymbol{\Omega}(\mathbf{c}^-, \mathbf{b}^+) = 0, \tag{9.51}$$

$$\boldsymbol{\Omega}(\mathbf{c}^-, \mathbf{c}^+) = \frac{15(\mathcal{P}_{\mathbf{c}})^2}{4\alpha^2}\big((\psi_{\mathbf{c}}^-)_{\xi}\psi_{\mathbf{c}}^+ - \psi_{\mathbf{c}}^-(\psi_{\mathbf{c}}^+)_{\xi}\big). \tag{9.52}$$

Since the only non-vanishing terms are on the diagonal of the matrix we can rewrite Π as

$$\Pi = \boldsymbol{\Omega}(\mathbf{a}^-, \mathbf{a}^+)\boldsymbol{\Omega}(\mathbf{b}^-, \mathbf{b}^+)\boldsymbol{\Omega}(\mathbf{c}^-, \mathbf{c}^+).$$

By using the Lagrangian subspace property of the stable and unstable subspaces, it follows that each component of this expression is independent of ξ and so they can be evaluated at any value of ξ. Now since

$$(\psi_{\mathbf{a}}^{\pm})_\xi = \alpha e^{\mp\alpha\kappa_1\xi}\,\mathrm{sech}^2(\alpha\xi)\left[1+2p\pm 2\kappa_1\tanh(\alpha\xi)+3\tanh^2(\alpha\xi)\right]\mp\alpha\kappa_1\psi_{\mathbf{a}}^{\pm},$$
$$(\psi_{\mathbf{b}}^{\pm})_\xi = \alpha e^{\mp\alpha\kappa_2\xi}\,\mathrm{sech}^2(\alpha\xi)\left[1+4p\pm 2\kappa_2\tanh(\alpha\xi)+3\tanh^2(\alpha\xi)\right]\mp\alpha\kappa_2\psi_{\mathbf{b}}^{\pm},$$
$$(\psi_{\mathbf{c}}^{\pm})_\xi = \alpha e^{\mp\alpha\kappa_3\xi}\,\mathrm{sech}^2(\alpha\xi)\left[1+6p\pm 2\kappa_3\tanh(\alpha\xi)+3\tanh^2(\alpha\xi)\right]\mp\alpha\kappa_3\psi_{\mathbf{c}}^{\pm},$$

evaluating at $\xi = 0$ will yield

$$\begin{aligned}
\psi_{\mathbf{a}}^{\pm}(0) &= \pm\frac{p\kappa_1}{3},\\
(\psi_{\mathbf{a}}^{\pm})_\xi(0) &= \alpha\left(1+2p-\frac{p(4+5p)}{3}\right),\\
\psi_{\mathbf{b}}^{\pm}(0) &= \pm\frac{2p\kappa_2}{3},\\
(\psi_{\mathbf{b}}^{\pm})_\xi(0) &= \alpha\left(1+4p-\frac{2p(4+10p)}{3}\right),\\
\psi_{\mathbf{c}}^{\pm}(0) &= \pm p\kappa_3,\\
(\psi_{\mathbf{c}}^{\pm})_\xi(0) &= \alpha\left(1+6p-p(4+15p)\right),
\end{aligned}$$

therefore

$$\begin{aligned}
\mathbf{\Omega}(\mathbf{a}^-,\mathbf{a}^+) &= \frac{15(\mathcal{P}_{\mathbf{a}})^2}{4\alpha}\left[\left(1+2p-\frac{4p+5p^2}{3}\right)\left(\frac{p\kappa_1}{3}\right)\right.\\
&\qquad\left.-\left(-\frac{p\kappa_1}{3}\right)\left(1+2p-\frac{4p+5p^2}{3}\right)\right]\\
&= \frac{5p\kappa_1(\mathcal{P}_{\mathbf{a}})^2}{2\alpha}\left(1+2p-\frac{4p+5p^2}{3}\right)\\
&= \frac{5p\kappa_1(\mathcal{P}_{\mathbf{a}})^2}{6\alpha}(3+2p-5p^2)\\
&= \frac{5p(\mathcal{P}_{\mathbf{a}})^2}{6\alpha}\sqrt{4+5p}\,(1-p)(3+5p), \qquad (9.53)\\
\mathbf{\Omega}(\mathbf{b}^-,\mathbf{b}^+) &= \frac{15(\mathcal{P}_{\mathbf{b}})^2}{4\alpha}\left[\left(1+4p-\frac{8p+20p^2}{3}\right)\left(\frac{2p\kappa_2}{3}\right)\right.\\
&\qquad\left.-\left(-\frac{2p\kappa_2}{3}\right)\left(1+4p-\frac{8p+20p^2}{3}\right)\right]
\end{aligned}$$

$$
\begin{aligned}
&= \frac{5p\kappa_2(\mathcal{P}_{\mathbf{b}})^2}{\alpha}\left(1+4p-\frac{8p+20p^2}{3}\right) \\
&= \frac{5p\kappa_2(\mathcal{P}_{\mathbf{b}})^2}{3\alpha}(3+4p-20p^2) \\
&= \frac{5p(\mathcal{P}_{\mathbf{b}})^2}{3\alpha}\sqrt{4+10p}\,(1-2p)(3+10p)\,, \qquad (9.54)
\end{aligned}
$$

$$
\begin{aligned}
\boldsymbol{\Omega}(\mathbf{c}^-,\mathbf{c}^+) &= \frac{15(\mathcal{P}_{\mathbf{c}})^2}{4\alpha}\left[(1+2p-15p^2)(p\kappa_3)-(-p\kappa_3)(1+2p-15p^2)\right] \\
&= \frac{15p\kappa_3(\mathcal{P}_{\mathbf{c}})^2}{2\alpha}(1+2p-15p^2) \\
&= \frac{15p(\mathcal{P}_{\mathbf{c}})^2}{2\alpha}\sqrt{4+15p}\,(1-3p)(1+5p)\,. \qquad (9.55)
\end{aligned}
$$

Since

$$
(\mathcal{P}_{\mathbf{a}})^2(\mathcal{P}_{\mathbf{b}})^2(\mathcal{P}_{\mathbf{c}})^2 = 81(\sigma_1)^2(\sigma_2)^2(\sigma_3)^2\widehat{\mathcal{P}} = \frac{24\alpha^3}{125\kappa_1\kappa_2\kappa_3}\widehat{\mathcal{P}}\,,
$$

where

$$
\begin{aligned}
\widehat{\mathcal{P}} = \Big[&\big((p+3)\sqrt{4+5p}+6p+6\big)\big((2p+3)\sqrt{4+10p}+12p+6\big) \\
&\times \big((p+1)\sqrt{4+15p}+6p+2\big)\Big]^{-2},
\end{aligned}
$$

by combining the above expressions the Lazutkin–Treschev invariant is found to be:

$$
\Pi = 2p^3(1-p)(1-2p)(1-3p)(3+5p)(3+10p)(1+5p)\widehat{\mathcal{P}}\,. \qquad (9.56)
$$

Hence

$$
\Pi \neq 0 \quad \text{when} \quad p \neq \left\{-\frac{3}{5}, -\frac{3}{10}, -\frac{1}{5}, 0, \frac{1}{3}, \frac{1}{2}, 1\right\}.
$$

9.5 The Remaining Hypotheses

It remains to check hypotheses (**H6**) and (**H7**). The λ, ν-dependent eigenvalues are

$$
\mu_1(\lambda,\nu) = \alpha^2\Upsilon - \alpha\sqrt{\widehat{\Upsilon}+4+15p}\,,
$$

$$\mu_2(\lambda,\nu) = \alpha^2\Upsilon - \alpha\sqrt{\widehat{\Upsilon}+4+10p}\,,$$

$$\mu_3(\lambda,\nu) = \alpha^2\Upsilon - \alpha\sqrt{\widehat{\Upsilon}+4+5p}\,,$$

$$\mu_4(\lambda,\nu) = \alpha^2\Upsilon - \alpha\sqrt{\widehat{\Upsilon}+4}\,,$$

$$\mu_5(\lambda,\nu) = \alpha^2\Upsilon + \alpha\sqrt{\widehat{\Upsilon}+4}\,,$$

$$\mu_6(\lambda,\nu) = \alpha^2\Upsilon + \alpha\sqrt{\widehat{\Upsilon}+4+5p}\,,$$

$$\mu_7(\lambda,\nu) = \alpha^2\Upsilon + \alpha\sqrt{\widehat{\Upsilon}+4+10p}\,,$$

$$\mu_8(\lambda,\nu) = \alpha^2\Upsilon + \alpha\sqrt{\widehat{\Upsilon}+4+15p}\,,$$

where Υ is given above and

$$\begin{aligned}\widehat{\Upsilon} &= \alpha^2\Upsilon^2 + \lambda^2 + \nu^2\\ &= (1+\ell^2)\lambda^2 + (1-c^2)\nu^2 - 2\mathtt{i}c\ell\lambda\nu\,.\end{aligned}$$

These eight eigenvalues are simple and form a four–four splitting for all λ in a neighbourhood of the real axis (our area of interest). Hence, with a suitable definition of Λ, Hypothesis (**H6**) will be satisfied.

Secondly, the exact solutions to (9.36)–(9.39) are given by

$$\begin{aligned}\psi_1^{\pm}(\xi,\lambda,\nu) &= e^{\mp\alpha\gamma_1\xi}A^{\pm}(\xi,\lambda,\nu)\,,\\ \psi_2^{\pm}(\xi,\lambda,\nu) &= e^{\mp\alpha\gamma_2\xi}B^{\pm}(\xi,\lambda,\nu)\,,\\ \psi_3^{\pm}(\xi,\lambda,\nu) &= e^{\mp\alpha\gamma_3\xi}C^{\pm}(\xi,\lambda,\nu)\,,\\ \psi_4^{\pm}(\xi,\lambda,\nu) &= e^{\mp\alpha\gamma_4\xi}D^{\pm}(\xi,\lambda,\nu)\,,\end{aligned}$$

where

$$\gamma_1(\lambda) = \sqrt{4+\widehat{\Upsilon}^2}\,, \qquad \gamma_2(\lambda) = \sqrt{4+5p+\widehat{\Upsilon}^2}\,,$$

$$\gamma_3(\lambda) = \sqrt{4+10p+\widehat{\Upsilon}^2}\,, \qquad \gamma_4(\lambda) = \sqrt{4+15p+\widehat{\Upsilon}^2}\,,$$

and

$$\begin{aligned}A^{\pm}(\xi,\lambda,\nu) = \pm\frac{\gamma_1\widehat{\Upsilon}}{15} + \left(1+\frac{2\widehat{\Upsilon}}{5}\right)\tanh(\alpha\xi) \pm \gamma_1\tanh^2(\alpha\xi)\\ + \tanh^3(\alpha\xi)\,,\end{aligned}$$

$$
\begin{aligned}
B^{\pm}(\xi,\lambda,\nu) &= \pm\frac{\gamma_2(5p+\widehat{\Upsilon})}{15} + \left(1+2p+\frac{2\widehat{\Upsilon}}{5}\right)\tanh(\alpha\xi) \pm \gamma_2\tanh^2(\alpha\xi) \\
&\quad + \tanh^3(\alpha\xi)\,, \\
C^{\pm}(\xi,\lambda,\nu) &= \pm\frac{\gamma_3(10p+\widehat{\Upsilon})}{15} + \left(1+4p+\frac{2\widehat{\Upsilon}}{5}\right)\tanh(\alpha\xi) \pm \gamma_3\tanh^2(\alpha\xi) \\
&\quad + \tanh^3(\alpha\xi)\,, \\
D^{\pm}(\xi,\lambda,\nu) &= \pm\frac{\gamma_4(15p+\widehat{\Upsilon})}{15} + \left(1+6p+\frac{2\widehat{\Upsilon}}{5}\right)\tanh(\alpha\xi) \pm \gamma_4\tanh^2(\alpha\xi) \\
&\quad + \tanh^3(\alpha\xi).
\end{aligned}
$$

Note that

$$
\begin{aligned}
&\psi_1^{\pm}(\xi,0,0) = -\frac{1}{2\alpha}\hat{\phi}_\xi(\xi)\,, \quad \psi_2^{\pm}(\xi,0,0) = \psi_{\mathbf{a}}^{\pm}(\xi)\,, \quad \psi_3^{\pm}(\xi,0,0) = \psi_{\mathbf{b}}^{\pm}(\xi)\,, \\
&\psi_4^{\pm}(\xi,0,0) = \psi_{\mathbf{c}}^{\pm}(\xi)\,.
\end{aligned} \tag{9.57}
$$

Reversing the transformations allows us to write linearly independent solutions to (9.15) as

$$
Z_1(\xi,\lambda,\nu) = e^{\mu_1(\lambda,\nu)\xi}\begin{pmatrix}
\frac{3}{2}D^+ \\
[\frac{3}{2}(c\mu_1+\lambda)+3(\ell\mu_1+\mathrm{i}\nu)-\mu_1]D^+ + [\frac{3}{2}c+3\ell-1]D_\xi^+ \\
[\frac{3}{2}\mu_1-(c\mu_1+\lambda)-2(\ell\mu_1+\mathrm{i}\nu)]D^+ + [\frac{3}{2}-c-2\ell]D_\xi^+ \\
[2\mu_1+3(c\mu_1+\lambda)+\frac{3}{2}(\ell\mu_1+\mathrm{i}\nu)]D^+ + [2+3c+\frac{3}{2}\ell]D_\xi^+ \\
2D^+ \\
3D^+ \\
D^+ \\
[3\mu_1+2(c\mu_1+\lambda)+(\ell\mu_1+\mathrm{i}\nu)]D^+ + [3+2c+\ell]D_\xi^+
\end{pmatrix},
$$

$$
Z_2(\xi,\lambda,\nu) = e^{\mu_2(\lambda,\nu)\xi}\begin{pmatrix}
2C^+ \\
[2(c\mu_2+\lambda)+(\ell\mu_2+\mathrm{i}\nu)-3\mu_2]C^+ + [2c+\ell-3]C_\xi^+ \\
[2\mu_2-3(c\mu_2+\lambda)-\frac{3}{2}(\ell\mu_2+\mathrm{i}\nu)]C^+ + [2-3c-\frac{3}{2}\ell]C_\xi^+ \\
[\frac{3}{2}\mu_2+(c\mu_2+\lambda)+2(\ell\mu_2+\mathrm{i}\nu)]C^+ + [\frac{3}{2}+c+2\ell]C_\xi^+ \\
\frac{3}{2}C^+ \\
C^+ \\
3C^+ \\
[\mu_2+\frac{3}{2}(c\mu_2+\lambda)+3(\ell\mu_2+\mathrm{i}\nu)]C^+ + [1+\frac{3}{2}c+3\ell]C_\xi^+
\end{pmatrix},
$$

$$Z_3(\xi,\lambda,\nu)=e^{\mu_3(\lambda,\nu)\xi}\begin{pmatrix}3B^+\\ [3(c\mu_3+\lambda)+\frac{3}{2}(\ell\mu_3+\mathtt{i}\nu)-2\mu_3]B^+ +[3c+\frac{3}{2}\ell-2]B^+_\xi\\ [3\mu_3-2(c\mu_3+\lambda)-(\ell\mu_3+\mathtt{i}\nu)]B^+ +[3-2c-\ell]B^+_\xi\\ [\mu_3+\frac{3}{2}(c\mu_3+\lambda)+3(\ell\mu_3+\mathtt{i}\nu)]B^+ +[1+\frac{3}{2}c+3\ell]B^+_\xi\\ B^+\\ \frac{3}{2}B^+\\ 2B^+\\ [\frac{3}{2}\mu_3+(c\mu_3+\lambda)+2(\ell\mu_3+\mathtt{i}\nu)]B^+ +[\frac{3}{2}+c+2\ell]B^+_\xi\end{pmatrix},$$

$$Z_4(\xi,\lambda,\nu)=e^{\mu_4(\lambda,\nu)\xi}\begin{pmatrix}A^+\\ [(c\mu_4+\lambda)+2(\ell\mu_4+\mathtt{i}\nu)-\frac{3}{2}\mu_4]A^+ +[c+2\ell-\frac{3}{2}]A^+_\xi\\ [\mu_4-\frac{3}{2}(c\mu_4+\lambda)-3(\ell\mu_4+\mathtt{i}\nu)]A^+ +[1-\frac{3}{2}c-3\ell]A^+_\xi\\ [3\mu_4+2(c\mu_4+\lambda)+(\ell\mu_4+\mathtt{i}\nu)]A^+ +[3+2c+\ell]A^+_\xi\\ 3A^+\\ 2A^+\\ \frac{3}{2}A^+\\ [2\mu_4+3(c\mu_4+\lambda)+\frac{3}{2}(\ell\mu_4+\mathtt{i}\nu)]A^+ +[2+3c+\frac{3}{2}\ell]A^+_\xi\end{pmatrix},$$

$$Z_5(\xi,\lambda,\nu)=e^{\mu_5(\lambda,\nu)\xi}\begin{pmatrix}A^-\\ [(c\mu_5+\lambda)+2(\ell\mu_5+\mathtt{i}\nu)-\frac{3}{2}\mu_5]A^- +[c+2\ell-\frac{3}{2}]A^-_\xi\\ [\mu_5-\frac{3}{2}(c\mu_5+\lambda)-3(\ell\mu_5+\mathtt{i}\nu)]A^- +[1-\frac{3}{2}c-3\ell]A^-_\xi\\ [3\mu_5+2(c\mu_5+\lambda)+(\ell\mu_5+\mathtt{i}\nu)]A^- +[3+2c+\ell]A^-_\xi\\ 3A^-\\ 2A^-\\ \frac{3}{2}A^-\\ [2\mu_5+3(c\mu_5+\lambda)+\frac{3}{2}(\ell\mu_5+\mathtt{i}\nu)]A^- +[2+3c+\frac{3}{2}\ell]A^-_\xi\end{pmatrix},$$

$$Z_6(\xi,\lambda,\nu)=e^{\mu_6(\lambda,\nu)\xi}\begin{pmatrix}3B^-\\ [3(c\mu_6+\lambda)+\frac{3}{2}(\ell\mu_6+\mathtt{i}\nu)-2\mu_6]B^- +[3c+\frac{3}{2}\ell-2]B^-_\xi\\ [3\mu_6-2(c\mu_6+\lambda)-(\ell\mu_6+\mathtt{i}\nu)]B^- +[3-2c-\ell]B^-_\xi\\ [\mu_6+\frac{3}{2}(c\mu_6+\lambda)+3(\ell\mu_6+\mathtt{i}\nu)]B^- +[1+\frac{3}{2}c+3\ell]B^-_\xi\\ B^-\\ \frac{3}{2}B^-\\ 2B^-\\ [\frac{3}{2}\mu_6+(c\mu_6+\lambda)+2(\ell\mu_6+\mathtt{i}\nu)]B^- +[\frac{3}{2}+c+2\ell]B^-_\xi\end{pmatrix},$$

$$Z_7(\xi,\lambda,\nu)=e^{\mu_7(\lambda,\nu)\xi}\begin{pmatrix} 2C^- \\ [2(c\mu_7+\lambda)+(\ell\mu_7+\mathrm{i}\nu)-3\mu_7]C^-+[2c+\ell-3]C^-_\xi \\ [2\mu_7-3(c\mu_7+\lambda)-\frac{3}{2}(\ell\mu_7+\mathrm{i}\nu)]C^-+[2-3c-\frac{3}{2}\ell]C^-_\xi \\ [\frac{3}{2}\mu_7+(c\mu_7+\lambda)+2(\ell\mu_7+\mathrm{i}\nu)]C^-+[\frac{3}{2}+c+2\ell]C^-_\xi \\ \frac{3}{2}C^- \\ C^- \\ 3C^- \\ [\mu_7+\frac{3}{2}(c\mu_7+\lambda)+3(\ell\mu_7+\mathrm{i}\nu)]C^-+[1+\frac{3}{2}c+3\ell]C^-_\xi \end{pmatrix},$$

$$Z_8(\xi,\lambda,\nu)=e^{\mu_8(\lambda,\nu)\xi}\begin{pmatrix} \frac{3}{2}D^- \\ [\frac{3}{2}(c\mu_8+\lambda)+3(\ell\mu_8+\mathrm{i}\nu)-\mu_8]D^-+[\frac{3}{2}c+3\ell-1]D^-_\xi \\ [\frac{3}{2}\mu_8-(c\mu_8+\lambda)-2(\ell\mu_8+\mathrm{i}\nu)]D^-+[\frac{3}{2}-c-2\ell]D^-_\xi \\ [2\mu_8+3(c\mu_8+\lambda)+\frac{3}{2}(\ell\mu_8+\mathrm{i}\nu)]D^-+[2+3c+\frac{3}{2}\ell]D^-_\xi \\ 2D^- \\ 3D^- \\ D^- \\ [3\mu_8+2(c\mu_8+\lambda)+(\ell\mu_8+\mathrm{i}\nu)]D^-+[3+2c+\ell]D^-_\xi \end{pmatrix}.$$

9.6 The Adjoint Equation

The solutions to the adjoint equation

$$\mathbf{J}W_\xi=[\mathbf{B}(\xi,c)+\lambda\mathbf{M}+\mathrm{i}\nu\mathbf{N}]\,W,\quad W\in\mathbb{C}^8 \tag{9.58}$$

can be obtained by replacing λ with $-\lambda$ and ν with $-\nu$ in the Z_j's above to get

$$W_1(\xi,\lambda,\nu)=e^{-\mu_1(\lambda,\nu)\xi}\begin{pmatrix} \frac{3}{2}D^- \\ [\frac{3}{2}c+3\ell-1]D^-_\xi-[\frac{3}{2}(c\mu_1+\lambda)+3(\ell\mu_1+\mathrm{i}\nu)-\mu_1]D^- \\ [\frac{3}{2}-c-2\ell]D^-_\xi-[\frac{3}{2}\mu_1-(c\mu_1+\lambda)-2(\ell\mu_1+\mathrm{i}\nu)]D^- \\ [2+3c+\frac{3}{2}\ell]D^-_\xi-[2\mu_1+3(c\mu_1+\lambda)+\frac{3}{2}(\ell\mu_1+\mathrm{i}\nu)]D^- \\ 2D^- \\ 3D^- \\ D^- \\ [3+2c+\ell]D^-_\xi-[3\mu_1+2(c\mu_1+\lambda)+(\ell\mu_1+\mathrm{i}\nu)]D^- \end{pmatrix},$$

$$W_2(\xi,\lambda,\nu)=e^{-\mu_2(\lambda,\nu)\xi}\begin{pmatrix} 2C^- \\ [2c+\ell-3]C^-_\xi-[2(c\mu_2+\lambda)+(\ell\mu_2+\mathrm{i}\nu)-3\mu_2]C^- \\ [2-3c-\frac{3}{2}\ell]C^-_\xi-[2\mu_2-3(c\mu_2+\lambda)-\frac{3}{2}(\ell\mu_2+\mathrm{i}\nu)]C^- \\ [\frac{3}{2}+c+2\ell]C^-_\xi-[\frac{3}{2}\mu_2+(c\mu_2+\lambda)+2(\ell\mu_2+\mathrm{i}\nu)]C^- \\ \frac{3}{2}C^- \\ C^- \\ 3C^- \\ [1+\frac{3}{2}c+3\ell]C^-_\xi-[\mu_2+\frac{3}{2}(c\mu_2+\lambda)+3(\ell\mu_2+\mathrm{i}\nu)]C^- \end{pmatrix},$$

$$W_3(\xi,\lambda,\nu)=e^{-\mu_3(\lambda,\nu)\xi}\begin{pmatrix} 3B^- \\ [3c+\frac{3}{2}\ell-2]B^-_\xi-[3(c\mu_3+\lambda)+\frac{3}{2}(\ell\mu_3+\mathrm{i}\nu)-2\mu_3]B^- \\ [3-2c-\ell]B^-_\xi-[3\mu_3-2(c\mu_3+\lambda)-(\ell\mu_3+\mathrm{i}\nu)]B^- \\ [1+\frac{3}{2}c+3\ell]B^-_\xi-[\mu_3+\frac{3}{2}(c\mu_3+\lambda)+3(\ell\mu_3+\mathrm{i}\nu)]B^- \\ B^- \\ \frac{3}{2}B^- \\ 2B^- \\ [\frac{3}{2}+c+2\ell]B^-_\xi-[\frac{3}{2}\mu_3+(c\mu_3+\lambda)+2(\ell\mu_3+\mathrm{i}\nu)]B^- \end{pmatrix},$$

$$W_4(\xi,\lambda,\nu)=e^{-\mu_4(\lambda,\nu)\xi}\begin{pmatrix} A^- \\ [c+2\ell-\frac{3}{2}]A^-_\xi-[(c\mu_4+\lambda)+2(\ell\mu_4+\mathrm{i}\nu)-\frac{3}{2}\mu_4]A^- \\ [1-\frac{3}{2}c-3\ell]A^-_\xi-[\mu_4-\frac{3}{2}(c\mu_4+\lambda)-3(\ell\mu_4+\mathrm{i}\nu)]A^- \\ [3+2c+\ell]A^-_\xi-[3\mu_4+2(c\mu_4+\lambda)+(\ell\mu_4+\mathrm{i}\nu)]A^- \\ 3A^- \\ 2A^- \\ \frac{3}{2}A^- \\ [2+3c+\frac{3}{2}\ell]A^-_\xi-[2\mu_4+3(c\mu_4+\lambda)+\frac{3}{2}(\ell\mu_4+\mathrm{i}\nu)]A^- \end{pmatrix},$$

$$W_5(\xi,\lambda,\nu)=e^{-\mu_5(\lambda,\nu)\xi}\begin{pmatrix} A^+ \\ [c+2\ell-\frac{3}{2}]A^+_\xi-[(c\mu_5+\lambda)+2(\ell\mu_5+\mathrm{i}\nu)-\frac{3}{2}\mu_5]A^+ \\ [1-\frac{3}{2}c-3\ell]A^+_\xi-[\mu_5-\frac{3}{2}(c\mu_5+\lambda)-3(\ell\mu_5+\mathrm{i}\nu)]A^+ \\ [3+2c+\ell]A^+_\xi-[3\mu_5+2(c\mu_5+\lambda)+(\ell\mu_5+\mathrm{i}\nu)]A^+ \\ 3A^+ \\ 2A^+ \\ \frac{3}{2}A^+ \\ [2+3c+\frac{3}{2}\ell]A^+_\xi-[2\mu_5+3(c\mu_5+\lambda)+\frac{3}{2}(\ell\mu_5+\mathrm{i}\nu)]A^+ \end{pmatrix},$$

$$W_6(\xi,\lambda,\nu) = e^{-\mu_6(\lambda,\nu)\xi}\begin{pmatrix} 3B^+ \\ [3c+\frac{3}{2}\ell-2]B^+_\xi - [3(c\mu_6+\lambda)+\frac{3}{2}(\ell\mu_6+\mathrm{i}\nu)-2\mu_6]B^+ \\ [3-2c-\ell]B^+_\xi - [3\mu_6-2(c\mu_6+\lambda)-(\ell\mu_6+\mathrm{i}\nu)]B^+ \\ [1+\frac{3}{2}c+3\ell]B^+_\xi - [\mu_6+\frac{3}{2}(c\mu_6+\lambda)+3(\ell\mu_6+\mathrm{i}\nu)]B^+ \\ B^+ \\ \frac{3}{2}B^+ \\ 2B^+ \\ [\frac{3}{2}+c+2\ell]B^+_\xi - [\frac{3}{2}\mu_6+(c\mu_6+\lambda)+2(\ell\mu_6+\mathrm{i}\nu)]B^+ \end{pmatrix},$$

$$W_7(\xi,\lambda,\nu) = e^{-\mu_7(\lambda,\nu)\xi}\begin{pmatrix} 2C^+ \\ [2c+\ell-3]C^+_\xi - [2(c\mu_7+\lambda)+(\ell\mu_7+\mathrm{i}\nu)-3\mu_7]C^+ \\ [2-3c-\frac{3}{2}\ell]C^+_\xi - [2\mu_7-3(c\mu_7+\lambda)-\frac{3}{2}(\ell\mu_7+\mathrm{i}\nu)]C^+ \\ [\frac{3}{2}+c+2\ell]C^+_\xi - [\frac{3}{2}\mu_7+(c\mu_7+\lambda)+2(\ell\mu_7+\mathrm{i}\nu)]C^+ \\ \frac{3}{2}C^+ \\ C^+ \\ 3C^+ \\ [1+\frac{3}{2}c+3\ell]C^+_\xi - [\mu_7+\frac{3}{2}(c\mu_7+\lambda)+3(\ell\mu_7+\mathrm{i}\nu)]C^+ \end{pmatrix},$$

$$W_8(\xi,\lambda,\nu) = e^{-\mu_8(\lambda,\nu)\xi}\begin{pmatrix} \frac{3}{2}D^+ \\ [\frac{3}{2}c+3\ell-1]D^+_\xi - [\frac{3}{2}(c\mu_8+\lambda)+3(\ell\mu_8+\mathrm{i}\nu)-\mu_8]D^+ \\ [\frac{3}{2}-c-2\ell]D^+_\xi - [\frac{3}{2}\mu_8-(c\mu_8+\lambda)-2(\ell\mu_8+\mathrm{i}\nu)]D^+ \\ [2+3c+\frac{3}{2}\ell]D^+_\xi - [2\mu_8+3(c\mu_8+\lambda)+\frac{3}{2}(\ell\mu_8+\mathrm{i}\nu)]D^+ \\ 2D^+ \\ 3D^+ \\ D^+ \\ [3+2c+\ell]D^+_\xi - [3\mu_8+2(c\mu_8+\lambda)+(\ell\mu_8+\mathrm{i}\nu)]D^+ \end{pmatrix}.$$

using $\mu_j(-\lambda,-\nu) = -\mu_{(9-j)}(\lambda,\nu)$ and

$$A^\pm(\xi,\lambda,\nu) = A^\pm(\xi,-\lambda,-\nu)\,, \qquad B^\pm(\xi,\lambda,\nu) = B^\pm(\xi,-\lambda,-\nu)\,,$$

$$C^\pm(\xi,\lambda,\nu) = C^\pm(\xi,-\lambda,-\nu)\,, \qquad D^\pm(\xi,\lambda,\nu) = D^\pm(\xi,-\lambda,-\nu)\,.$$

The λ, ν-dependent eigenvectors and adjoint eigenvectors are

$$\zeta_1(\lambda,\nu)=\sigma_1(\lambda,\nu)\begin{pmatrix}\frac{3}{2}\\ \frac{3}{2}(c\mu_1+\lambda)+3(\ell\mu_1+\mathrm{i}\nu)-\mu_1\\ \frac{3}{2}\mu_1-(c\mu_1+\lambda)-2(\ell\mu_1+\mathrm{i}\nu)\\ 2\mu_1+3(c\mu_1+\lambda)+\frac{3}{2}(\ell\mu_1+\mathrm{i}\nu)\\ 2\\ 3\\ 1\\ 3\mu_1+2(c\mu_1+\lambda)+(\ell\mu_1+\mathrm{i}\nu)\end{pmatrix},$$

$$\zeta_2(\lambda,\nu)=\sigma_2(\lambda,\nu)\begin{pmatrix}2\\ 2(c\mu_2+\lambda)+(\ell\mu_2+\mathrm{i}\nu)-3\mu_2\\ 2\mu_2-3(c\mu_2+\lambda)-\frac{3}{2}(\ell\mu_2+\mathrm{i}\nu)\\ \frac{3}{2}\mu_2+(c\mu_2+\lambda)+2(\ell\mu_2+\mathrm{i}\nu)\\ \frac{3}{2}\\ 1\\ 3\\ \mu_2+\frac{3}{2}(c\mu_2+\lambda)+3(\ell\mu_2+\mathrm{i}\nu)\end{pmatrix},$$

$$\zeta_3(\lambda,\nu)=\sigma_3(\lambda,\nu)\begin{pmatrix}3\\ 3(c\mu_3+\lambda)+\frac{3}{2}(\ell\mu_3+\mathrm{i}\nu)-2\mu_3\\ 3\mu_3-2(c\mu_3+\lambda)-(\ell\mu_3+\mathrm{i}\nu)\\ \mu_3+\frac{3}{2}(c\mu_3+\lambda)+3(\ell\mu_3+\mathrm{i}\nu)\\ 1\\ \frac{3}{2}\\ 2\\ \frac{3}{2}\mu_3+(c\mu_3+\lambda)+2(\ell\mu_3+\mathrm{i}\nu)\end{pmatrix},$$

$$\zeta_4(\lambda,\nu)=\sigma_4(\lambda,\nu)\begin{pmatrix}1\\ (c\mu_4+\lambda)+2(\ell\mu_4+\mathrm{i}\nu)-\frac{3}{2}\mu_4\\ \mu_4-\frac{3}{2}(c\mu_4+\lambda)-3(\ell\mu_4+\mathrm{i}\nu)\\ 3\mu_4+2(c\mu_4+\lambda)+(\ell\mu_4+\mathrm{i}\nu)\\ 3\\ 2\\ \frac{3}{2}\\ 2\mu_4+3(c\mu_4+\lambda)+\frac{3}{2}(\ell\mu_4+\mathrm{i}\nu)\end{pmatrix},$$

$$\zeta_5(\lambda,\nu) = -\sigma_4(\lambda,\nu)\begin{pmatrix} 1 \\ (c\mu_5+\lambda)+2(\ell\mu_5+\mathrm{i}\nu)-\frac{3}{2}\mu_5 \\ \mu_5-\frac{3}{2}(c\mu_5+\lambda)-3(\ell\mu_5+\mathrm{i}\nu) \\ 3\mu_5+2(c\mu_5+\lambda)+(\ell\mu_5+\mathrm{i}\nu) \\ 3 \\ 2 \\ \frac{3}{2} \\ 2\mu_5+3(c\mu_5+\lambda)+\frac{3}{2}(\ell\mu_5+\mathrm{i}\nu) \end{pmatrix},$$

$$\zeta_6(\lambda,\nu) = -\sigma_3(\lambda,\nu)\begin{pmatrix} 3 \\ 3(c\mu_6+\lambda)+\frac{3}{2}(\ell\mu_6+\mathrm{i}\nu)-2\mu_6 \\ 3\mu_6-2(c\mu_6+\lambda)-(\ell\mu_6+\mathrm{i}\nu) \\ \mu_6+\frac{3}{2}(c\mu_6+\lambda)+3(\ell\mu_6+\mathrm{i}\nu) \\ 1 \\ \frac{3}{2} \\ 2 \\ \frac{3}{2}\mu_6+(c\mu_6+\lambda)+2(\ell\mu_6+\mathrm{i}\nu) \end{pmatrix},$$

$$\zeta_7(\lambda,\nu) = \sigma_2(\lambda,\nu)\begin{pmatrix} 2 \\ 2(c\mu_7+\lambda)+(\ell\mu_7+\mathrm{i}\nu)-3\mu_7 \\ 2\mu_7-3(c\mu_7+\lambda)-\frac{3}{2}(\ell\mu_7+\mathrm{i}\nu) \\ \frac{3}{2}\mu_7+(c\mu_7+\lambda)+2(\ell\mu_7+\mathrm{i}\nu) \\ \frac{3}{2} \\ 1 \\ 3 \\ \mu_7+\frac{3}{2}(c\mu_7+\lambda)+3(\ell\mu_7+\mathrm{i}\nu) \end{pmatrix},$$

$$\zeta_8(\lambda,\nu) = \sigma_1(\lambda,\nu)\begin{pmatrix} \frac{3}{2} \\ \frac{3}{2}(c\mu_8+\lambda)+3(\ell\mu_8+\mathrm{i}\nu)-\mu_8 \\ \frac{3}{2}\mu_8-(c\mu_8+\lambda)-2(\ell\mu_8+\mathrm{i}\nu) \\ 2\mu_8+3(c\mu_8+\lambda)+\frac{3}{2}(\ell\mu_8+\mathrm{i}\nu) \\ 2 \\ 3 \\ 1 \\ 3\mu_8+2(c\mu_8+\lambda)+(\ell\mu_8+\mathrm{i}\nu) \end{pmatrix}.$$

$$\eta_1(\lambda,\nu) = -\sigma_1(\lambda,\nu)\begin{pmatrix} \frac{3}{2} \\ \mu_1 - \frac{3}{2}(c\mu_1+\lambda) - 3(\ell\mu_1+\mathrm{i}\nu) \\ (c\mu_1+\lambda) + 2(\ell\mu_1+\mathrm{i}\nu) - \frac{3}{2}\mu_1 \\ -2\mu_1 - 3(c\mu_1+\lambda) - \frac{3}{2}(\ell\mu_1+\mathrm{i}\nu) \\ 2 \\ 3 \\ 1 \\ -3\mu_1 - 2(c\mu_1+\lambda) - (\ell\mu_1+\mathrm{i}\nu) \end{pmatrix},$$

$$\eta_2(\lambda,\nu) = -\sigma_2(\lambda,\nu)\begin{pmatrix} 2 \\ 3\mu_2 - 2(c\mu_2+\lambda) - (\ell\mu_2+\mathrm{i}\nu) \\ 3(c\mu_2+\lambda) + \frac{3}{2}(\ell\mu_2+\mathrm{i}\nu) - 2\mu_2 \\ -\frac{3}{2}\mu_2 - (c\mu_2+\lambda) - 2(\ell\mu_2+\mathrm{i}\nu) \\ \frac{3}{2} \\ 1 \\ 3 \\ -\mu_2 - \frac{3}{2}(c\mu_2+\lambda) - 3(\ell\mu_2+\mathrm{i}\nu) \end{pmatrix},$$

$$\eta_3(\lambda,\nu) = \sigma_3(\lambda,\nu)\begin{pmatrix} 3 \\ 2\mu_3 - 3(c\mu_3+\lambda) - \frac{3}{2}(\ell\mu_3+\mathrm{i}\nu) \\ 2(c\mu_3+\lambda) + (\ell\mu_3+\mathrm{i}\nu) - 3\mu_3 \\ -\mu_3 - \frac{3}{2}(c\mu_3+\lambda) - 3(\ell\mu_3+\mathrm{i}\nu) \\ 1 \\ \frac{3}{2} \\ 2 \\ -\frac{3}{2}\mu_3 - (c\mu_3+\lambda) - 2(\ell\mu_3+\mathrm{i}\nu) \end{pmatrix},$$

$$\eta_4(\lambda,\nu) = \sigma_4(\lambda,\nu)\begin{pmatrix} 1 \\ \frac{3}{2}\mu_4 - (c\mu_4+\lambda) - 2(\ell\mu_4+\mathrm{i}\nu) \\ \frac{3}{2}(c\mu_4+\lambda) + 3(\ell\mu_4+\mathrm{i}\nu) - \mu_4 \\ -3\mu_4 - 2(c\mu_4+\lambda) - (\ell\mu_4+\mathrm{i}\nu) \\ 3 \\ 2 \\ \frac{3}{2} \\ -2\mu_4 - 3(c\mu_4+\lambda) - \frac{3}{2}(\ell\mu_4+\mathrm{i}\nu) \end{pmatrix},$$

$$\eta_5(\lambda,\nu)=\sigma_4(\lambda,\nu)\begin{pmatrix}1\\ \frac{3}{2}\mu_5-(c\mu_5+\lambda)-2(\ell\mu_5+\mathrm{i}\nu)\\ \frac{3}{2}(c\mu_5+\lambda)+3(\ell\mu_5+\mathrm{i}\nu)-\mu_5\\ -3\mu_5-2(c\mu_5+\lambda)-(\ell\mu_5+\mathrm{i}\nu)\\ 3\\ 2\\ \frac{3}{2}\\ -2\mu_5-3(c\mu_5+\lambda)-\frac{3}{2}(\ell\mu_5+\mathrm{i}\nu)\end{pmatrix},$$

$$\eta_6(\lambda,\nu)=\sigma_3(\lambda,\nu)\begin{pmatrix}3\\ 2\mu_6-3(c\mu_6+\lambda)-\frac{3}{2}(\ell\mu_6+\mathrm{i}\nu)\\ 2(c\mu_6+\lambda)+(\ell\mu_6+\mathrm{i}\nu)-3\mu_6\\ -\mu_6-\frac{3}{2}(c\mu_6+\lambda)-3(\ell\mu_6+\mathrm{i}\nu)\\ 1\\ \frac{3}{2}\\ 2\\ -\frac{3}{2}\mu_6-(c\mu_6+\lambda)-2(\ell\mu_6+\mathrm{i}\nu)\end{pmatrix},$$

$$\eta_7(\lambda,\nu)=\sigma_2(\lambda,\nu)\begin{pmatrix}2\\ 3\mu_7-2(c\mu_7+\lambda)-(\ell\mu_7+\mathrm{i}\nu)\\ 3(c\mu_7+\lambda)+\frac{3}{2}(\ell\mu_7+\mathrm{i}\nu)-2\mu_7\\ -\frac{3}{2}\mu_7-(c\mu_7+\lambda)-2(\ell\mu_7+\mathrm{i}\nu)\\ \frac{3}{2}\\ 1\\ 3\\ -\mu_7-\frac{3}{2}(c\mu_7+\lambda)-3(\ell\mu_7+\mathrm{i}\nu)\end{pmatrix},$$

$$\eta_8(\lambda,\nu)=\sigma_1(\lambda,\nu)\begin{pmatrix}\frac{3}{2}\\ \mu_8-\frac{3}{2}(c\mu_8+\lambda)-3(\ell\mu_8+\mathrm{i}\nu)\\ (c\mu_8+\lambda)+2(\ell\mu_8+\mathrm{i}\nu)-\frac{3}{2}\mu_8\\ -2\mu_8-3(c\mu_8+\lambda)-\frac{3}{2}(\ell\mu_8+\mathrm{i}\nu)\\ 2\\ 3\\ 1\\ -3\mu_8-2(c\mu_8+\lambda)-(\ell\mu_8+\mathrm{i}\nu)\end{pmatrix}.$$

where

$$\sigma_1(\lambda,\nu)=\sqrt{\frac{2\alpha}{15\sqrt{\widehat{\Upsilon}}+4+15p}},\qquad \sigma_2(\lambda,\nu)=\sqrt{\frac{2\alpha}{15\sqrt{\widehat{\Upsilon}}+4+10p}},$$

$$\sigma_3(\lambda,\nu)=\sqrt{\frac{2\alpha}{15\sqrt{\widehat{\Upsilon}}+4+5p}},\qquad \sigma_4(\lambda,\nu)=\sqrt{\frac{2\alpha}{15\sqrt{\widehat{\Upsilon}}+4}},$$

are λ, ν-dependent extensions of the constants σ_1, σ_2, σ_3, σ_4. From this we can determine that the asymptotics of the above solutions are

$$\lim_{\xi\to+\infty} e^{-\mu_1\xi} Z_1(\xi,\lambda,\nu) = \frac{\tilde{D}(\lambda,\nu)}{\sigma_1(\lambda,\nu)}\zeta_1(\lambda,\nu)\,,$$

$$\lim_{\xi\to+\infty} e^{-\mu_2\xi} Z_2(\xi,\lambda,\nu) = \frac{\tilde{C}(\lambda,\nu)}{\sigma_2(\lambda,\nu)}\zeta_2(\lambda,\nu)\,,$$

$$\lim_{\xi\to+\infty} e^{-\mu_3\xi} Z_3(\xi,\lambda,\nu) = \frac{\tilde{B}(\lambda,\nu)}{\sigma_3(\lambda,\nu)}\zeta_3(\lambda,\nu)\,,$$

$$\lim_{\xi\to+\infty} e^{-\mu_4\xi} Z_4(\xi,\lambda,\nu) = \frac{\tilde{A}(\lambda,\nu)}{\sigma_4(\lambda,\nu)}\zeta_4(\lambda,\nu)\,,$$

$$\lim_{\xi\to-\infty} e^{-\mu_5\xi} Z_5(\xi,\lambda,\nu) = \frac{\tilde{A}(\lambda,\nu)}{\sigma_4(\lambda,\nu)}\zeta_5(\lambda,\nu)\,,$$

$$\lim_{\xi\to-\infty} e^{-\mu_6\xi} Z_6(\xi,\lambda,\nu) = \frac{\tilde{B}(\lambda,\nu)}{\sigma_3(\lambda,\nu)}\zeta_6(\lambda,\nu)\,,$$

$$\lim_{\xi\to-\infty} e^{-\mu_7\xi} Z_7(\xi,\lambda,\nu) = -\frac{\tilde{C}(\lambda,\nu)}{\sigma_2(\lambda,\nu)}\zeta_7(\lambda,\nu)\,,$$

$$\lim_{\xi\to-\infty} e^{-\mu_8\xi} Z_8(\xi,\lambda,\nu) = -\frac{\tilde{D}(\lambda,\nu)}{\sigma_1(\lambda,\nu)}\zeta_8(\lambda,\nu)\,,$$

$$\lim_{\xi\to-\infty} e^{+\mu_1\xi} W_1(\xi,\lambda,\nu) = \frac{\tilde{D}(\lambda,\nu)}{\sigma_1(\lambda,\nu)}\eta_1(\lambda,\nu)\,,$$

$$\lim_{\xi\to-\infty} e^{+\mu_2\xi} W_2(\xi,\lambda,\nu) = \frac{\tilde{C}(\lambda,\nu)}{\sigma_2(\lambda,\nu)}\eta_2(\lambda,\nu)\,,$$

$$\lim_{\xi\to-\infty} e^{+\mu_3\xi} W_3(\xi,\lambda,\nu) = -\frac{\tilde{B}(\lambda,\nu)}{\sigma_3(\lambda,\nu)}\eta_3(\lambda,\nu)\,,$$

$$\lim_{\xi\to-\infty} e^{+\mu_4\xi} W_4(\xi,\lambda,\nu) = -\frac{\tilde{A}(\lambda,\nu)}{\sigma_4(\lambda,\nu)}\eta_4(\lambda,\nu)\,,$$

$$\lim_{\xi\to+\infty} e^{+\mu_5\xi} W_5(\xi,\lambda,\nu) = \frac{\tilde{A}(\lambda,\nu)}{\sigma_4(\lambda,\nu)}\eta_5(\lambda,\nu)\,,$$

$$\lim_{\xi\to+\infty} e^{+\mu_6\xi} W_6(\xi,\lambda,\nu) = \frac{\tilde{B}(\lambda,\nu)}{\sigma_3(\lambda,\nu)}\eta_6(\lambda,\nu)\,,$$

$$\lim_{\xi\to+\infty} e^{+\mu_7\xi} W_7(\xi,\lambda,\nu) = \frac{\tilde{C}(\lambda,\nu)}{\sigma_2(\lambda,\nu)}\eta_7(\lambda,\nu)\,,$$

$$\lim_{\xi\to+\infty} e^{+\mu_8\xi} W_8(\xi,\lambda,\nu) = \frac{\tilde{D}(\lambda,\nu)}{\sigma_1(\lambda,\nu)}\eta_8(\lambda,\nu)\,,$$

where

$$\begin{aligned}
\tilde{A}(\lambda,\nu) &= 2 + \frac{2\widehat{\Upsilon}}{5} + \frac{(\widehat{\Upsilon}+15)\sqrt{\widehat{\Upsilon}+4}}{15}\,,\\
\tilde{B}(\lambda,\nu) &= 2 + 2p + \frac{2\widehat{\Upsilon}}{5} + \frac{(\widehat{\Upsilon}+5p+15)\sqrt{\widehat{\Upsilon}+4+5p}}{15}\,,\\
\tilde{C}(\lambda,\nu) &= 2 + 4p + \frac{2\widehat{\Upsilon}}{5} + \frac{(\widehat{\Upsilon}+10p+15)\sqrt{\widehat{\Upsilon}+4+10p}}{15}\,,\\
\tilde{D}(\lambda,\nu) &= 2 + 6p + \frac{2\widehat{\Upsilon}}{5} + \frac{(\widehat{\Upsilon}+15p+15)\sqrt{\widehat{\Upsilon}+4+15p}}{15}\,.
\end{aligned}$$

Hence, by taking

$$\begin{array}{llll}
\mathbf{u}_1 = \frac{\sigma_1}{\tilde{D}} Z_1\,, & \mathbf{u}_2 = \frac{\sigma_2}{\tilde{C}} Z_2\,, & \mathbf{u}_3 = \frac{\sigma_3}{\tilde{B}} Z_3\,, & \mathbf{u}_4 = \frac{\sigma_4}{\tilde{A}} Z_4\,,\\
\mathbf{u}_5 = \frac{\sigma_4}{\tilde{A}} Z_5\,, & \mathbf{u}_6 = \frac{\sigma_3}{\tilde{B}} Z_6\,, & \mathbf{u}_7 = -\frac{\sigma_2}{\tilde{C}} Z_7\,, & \mathbf{u}_8 = -\frac{\sigma_1}{\tilde{D}} Z_8\,,\\
\mathbf{w}_1 = \frac{\sigma_1}{\tilde{D}} W_1\,, & \mathbf{w}_2 = \frac{\sigma_2}{\tilde{C}} W_2\,, & \mathbf{w}_3 = -\frac{\sigma_3}{\tilde{B}} W_3\,, & \mathbf{w}_4 = -\frac{\sigma_4}{\tilde{A}} W_4\,,\\
\mathbf{w}_5 = \frac{\sigma_4}{\tilde{A}} W_5\,, & \mathbf{w}_6 = \frac{\sigma_3}{\tilde{B}} W_6\,, & \mathbf{w}_7 = \frac{\sigma_2}{\tilde{C}} W_7\,, & \mathbf{w}_8 = \frac{\sigma_1}{\tilde{D}} W_8\,,
\end{array}$$

the asymptotic conditions (6.13) and (6.19) are met. Next use (9.57) to deduce that

$$\begin{aligned}
Z_5(\xi,0,0) &= -\frac{1}{2\alpha}\widehat{Z}_\xi\,,\ Z_6(\xi,0,0) = \mathcal{P}_{\mathbf{a}}^{-1}\mathbf{a}^-\,,\\
Z_7(\xi,0,0) &= \mathcal{P}_{\mathbf{b}}^{-1}\mathbf{b}^-\,,\ Z_8(\xi,0,0) = \mathcal{P}_{\mathbf{c}}^{-1}\mathbf{c}^-\,,\\
W_5(\xi,0,0) &= -\frac{1}{2\alpha}\widehat{Z}_\xi\,,\ W_6(\xi,0,0) = \mathcal{P}_{\mathbf{a}}^{-1}\mathbf{a}^+\,,\\
W_7(\xi,0,0) &= \mathcal{P}_{\mathbf{b}}^{-1}\mathbf{b}^+\,,\ W_8(\xi,0,0) = \mathcal{P}_{\mathbf{c}}^{-1}\mathbf{c}^+\,,
\end{aligned} \tag{9.59}$$

which allows us to confirm (6.44) and (6.45):

$$\mathbf{u}_5(\xi,0,0)=\frac{1}{\chi^-}\widehat{Z}_\xi\,,\ \mathbf{u}_6(\xi,0,0)=\mathbf{a}^-\,,\ \mathbf{u}_7(\xi,0,0)=\mathbf{b}^-\,,\ \mathbf{u}_8(\xi,0,0)=\mathbf{c}^-\,,$$
$$\mathbf{w}_5(\xi,0,0)=\frac{1}{\chi^+}\widehat{Z}_\xi\,,\ \mathbf{w}_6(\xi,0,0)=\mathbf{a}^+\,,\ \mathbf{w}_7(\xi,0,0)=\mathbf{b}^+\,,\ \mathbf{w}_8(\xi,0,0)=\mathbf{c}^+\,, \tag{9.60}$$

using the values of χ^-, χ^+ from (9.24). Therefore, hypothesis (**H7**) is satisfied.

9.7 Summary

The key properties of the solitary wave that feed into the second derivatives of the Evans function are

$$\begin{aligned}
\chi &= \frac{1}{960\alpha} > 0,\\
\frac{\partial I_1}{\partial c} &= 4(1+\ell^2)\alpha^3 > 0,\\
\frac{\partial I_1}{\partial \ell} = \frac{\partial I_2}{\partial c} &= -4c\ell\alpha^3,\\
\frac{\partial I_2}{\partial \ell} &= 4(c^2-1)\alpha^3,\\
\Pi &= 2p^3(1-p)(1-2p)(1-3p)(3+5p)(3+10p)(1+5p)\widehat{\mathcal{P}},
\end{aligned}$$

and so

$$\begin{aligned}
D_{\lambda\lambda}(0,0) &= \frac{(1+\ell^2)\alpha^2}{240}\Pi,\\
D_{\lambda\nu}(0,0) &= -\mathrm{i}\frac{c\ell\alpha^2}{240}\Pi,\\
D_{\nu\nu}(0,0) &= \frac{(1-c^2)\alpha^2}{240}\Pi.
\end{aligned} \tag{9.61}$$

Now $\Pi > 0$ for p in the ranges $\left(0,\frac{1}{3}\right)$ and $\left(\frac{1}{2},1\right)$, while $\Pi < 0$ for p in the ranges $\left(\frac{1}{3},\frac{1}{2}\right)$ and $(1,\infty)$. This means that

$$p \in \left(0,\frac{1}{3}\right)\cup\left(\frac{1}{2},1\right) \quad \Rightarrow \quad D_{\lambda\lambda}(0,0),\, D_{\nu\nu}(0,0) > 0$$

and

$$p \in \left(\frac{1}{3}, \frac{1}{2}\right) \cup (1, \infty) \quad \Rightarrow \quad D_{\lambda\lambda}(0,0),\, D_{\nu\nu}(0,0) < 0$$

when $|c| < 1$. In the case where $1 < |c| < \sqrt{1 + \ell^2}$ this becomes

$$p \in \left(0, \frac{1}{3}\right) \cup \left(\frac{1}{2}, 1\right) \Rightarrow D_{\lambda\lambda}(0,0) > 0, \; D_{\nu\nu}(0,0) < 0,$$

$$p \in \left(\frac{1}{3}, \frac{1}{2}\right) \cup (1, \infty) \Rightarrow D_{\lambda\lambda}(0,0) < 0, \; D_{\nu\nu}(0,0) > 0,$$

while if $c = \pm 1$ then $D_{\nu\nu}(0,0) = 0$.

Now that the derivatives have been constructed, there are different ways that the leading-order Taylor expansion can be used to predict instability. The simplest case is to use Corollary 7.2, which states that positivity of the determinant (1.19) is a sufficient condition for instability. Evaluating the determinant,

$$\det \begin{bmatrix} \partial_c I_1 & \partial_\ell I_1 \\ \partial_c I_2 & \partial_\ell I_2 \end{bmatrix} = 16\alpha^6 (c^2 - \ell^2 - 1) = -16\alpha^4 \,.$$

Hence long-wave transverse instability does not occur. Another mechanism for instability is to look at $D(\lambda, \nu)$ along rays in the $(\mathrm{Re}(\lambda), \nu)$ plane, and look at the sign of $D(\lambda, \nu)$ along the ray in the infinite limit. Then check the sign of the derivative of D at the origin of the ray. In this case, since Π changes sign at several values of p, there will be regions of instability. Since this example was constructed for illustration only, we do not study these instability regions in detail.

Chapter 10
Concluding Remarks

The main accomplishments of this monograph are (a) that the theory is developed for an abstract class of Hamiltonian PDEs. Since almost all Hamiltonian PDEs known to the authors can be multisymplectified, this class of PDEs is quite general. (b) Alternatively, an entire abstract theory can be developed purely from geometry, building the PDEs starting from the geometry of space-time, and manifolds built on space-time. (c) The basic states are constructed based on a two-parameter variational principle for oblique line solitary waves, and the parameter structure of the variational principle feeds into the linear stability problem. (d) New algebraic structure, in the form of three independent symplectic forms, of the Evans function in $2+1$ is constructed. (e) A general instability condition is proved that uses only the geometric properties of the family of existing solitary wave. (f) The Maslov index arises in the linear instability condition in a surprising way. (g) An explicit example is constructed which illustrates the theory as well as the geometry of the linear stability problem.

The theory can be developed further in several directions. A direction of great interest is enlargement of the types of basic solitary waves. An intriguing example is the transverse instability of solitary waves, of the Kawahara–KP equation, where the states at infinity are distinct periodic travelling waves. The Kawahara–KP equation is

$$
\begin{aligned}
&u_t + f(u)_x + \alpha u_{xxx} + u_{xxxxx} + \sigma v_y = 0\,,\\
&v_x = u_y\,,
\end{aligned}
\tag{10.1}
$$

for scalar-valued functions $u(x, y, t)$ and $v(x, y, t)$, nonlinear function $f(u)$, and real parameters α and σ. The y-independent states satisfy the generalised Kawahara equation

$$
u_t + f(u)_x + \alpha u_{xxx} + u_{xxxxx} = 0\,.
$$

T. J. Burchell, T. J. Bridges, *Transverse Instability of Solitary Waves*, Mathematical Physics Studies, https://doi.org/10.1007/978-981-95-2046-6_10

This equation, in the case $f(u) = \frac{1}{2}u^2$, is studied by Sprenger et al. [134] and they find solitary waves that are biasymptotic to distinct periodic travelling waves of finite amplitude. Although it has not been shown heretofore, both the generalised Kawahara equation, and the Kawahara–KP equation are Hamiltonian PDEs and are amenable to being multisymplectified. The other facets of the theory of this monograph should generalise to this setting as well, but new challenges will arise.

A second direction of interest is allowing for greater generality in the position of the μ-eigenvalues of $\mathbf{A}^{\infty}(\lambda, \nu)$ in the assumption (**H6**). The assumption of simple eigenvalues is not too difficult to overcome. When the μ-eigenvalues are simple at $\lambda = 0$ they are simple for λ near zero and so, with the emphasis of the monograph on the derivatives of $D(\lambda, \nu)$ at $\lambda = 0$, the assumption of simple eigenvalues is adequate. However, for global (in λ) analysis, this assumption can be relaxed by working on exterior algebra spaces, where $\zeta_1 \wedge \cdots \wedge \zeta_4$ and $\zeta_5 \wedge \cdots \wedge \zeta_8$ can be constructed to be analytic for all $\lambda \in \Lambda$, or the problem of multiple eigenvalues can be addressed by constructing maximally-analytic individual eigenvectors and adjoint eigenvectors [31].

A bigger issue is the restriction of the μ-eigenvalues, in the case $\lambda = 0$, to be real. Complex eigenvalues can appear in quartets, when $\lambda = 0$, and are associated with an important class of solitary waves (an example being bright solitary waves). The question is whether the geometric instability criterion in Chap. 7 will be modified in some way. We suspect that it will be the same. In the $1 + 1$ case, the role of dI/dc in the stability of solitary waves is well established: implicitly in Grillakis et al. [72] and explicitly in Akmediev et al. [2] and Buryak and Akmediev [36]. The precise details of how to modify the proof of the derivative formula, for complex eigenvalues at $\lambda = 0$, in both the $1 + 1$ and $2 + 1$ cases, are an open question.

A third direction of interest is to expand the definition of the Evans function. In a seminal paper, Gesztesy et al. [69] show that the Evans function can be related, exactly, to an infinite-dimensional Fredholm determinant of a Birman–Schwinger-type integral operator, leading to an identity of the form

$$\det_2(I + K) = \mathrm{e}^{\theta} D\,, \tag{10.2}$$

where θ is a purely imaginary constant, D is the Evans function, and K is the Birman–Schwinger-type operator. This theory has been extended in several directions: extension to PDEs and connection with the Maslov index (e.g. Cox et al. [56]), and Das and Latushkin [58] use (10.2) to give a new derivative formula for the Evans function. It is of interest to generalise this formula to the case here where there is the triple symplectic structure of the Evans function ODEs. It is of interest moreover to use the symplectified (10.2) to give an alternative proof of the derivative formula in Chap. 7.

The assumption of an eight-dimensional phase space in the steady problem is sufficient to capture the essence of the theory in $2 + 1$, and avoids unnecessary complexity. When the phase space has dimension $2n$ with $n > 4$, the theory can in principle be generalised. The only change is that Π will have to be adapted to the new dimension of the stable and unstable manifolds. The 4–4 splitting would be replaced by an n–n splitting. The higher-dimensional case is better treated as it arises in applications.

One of the intriguing results that emerges in the theory is the role of the Maslov index in the derivative formula, which is connected to the parity–transversality formula (1.39). The proof of this formula relies on the Souriau construction of the Maslov index. It would be of interest to explore the role of the Maslov index, and its generalisations, in greater detail, particularly for more exotic solitary waves such as those of the Kawahara–KP equation mentioned above.

Appendix A
Multisymplectifying the KP Equation

In this appendix, a derivation of the multisymplectic structure for the KP equation (3.53) given in Sect. 3.8 is given. The starting point is the Lagrangian density

$$L = \frac{1}{2}\phi_{tx}\phi_{xx} + F\left(\phi_{xx}\right) - \frac{1}{2}\phi_{xxx}^2 + \frac{1}{2}\sigma\phi_{xy}^2 ,$$

for ϕ, where $\phi_{xx} = u$ with u satisfying the KP equation.

To implement the relaxed variational principle, introduce new coordinates

$$v = \phi_x , \quad u = \phi_{xx} , \quad w = \phi_{xxx} , \quad \text{and} \quad p = \phi_{xy} .$$

Substitute into the Lagrangian and add as constraints, enforced by Lagrange multipliers,

$$L = \frac{1}{2}uv_t + F(u) - \frac{1}{2}w^2 + \frac{1}{2}\sigma p^2 + \alpha(\phi_x - v) + \beta(v_x - u) + \gamma(u_x - w) + \tau(v_y - p) .$$

This Lagrangian now depends on $\phi, u, v, w, p, \alpha, \beta, \gamma$, and τ. Taking variations with respect to each coordinate, in the cube $\mathcal{C} := [x_1, x_2] \times [y_1, y_2] \times [t_1, t_2] \subset \mathbb{R}^3$ with fixed endpoint conditions gives

$$\begin{aligned}
\delta\phi &: -\alpha_x = 0 , \\
\delta v &: -\tfrac{1}{2}u_t - \alpha - \beta_x - \tau_y = 0 , \\
\delta u &: \tfrac{1}{2}v_t + f(u) - \beta - \gamma_x = 0 , \\
\delta w &: -w - \gamma = 0 , \\
\delta p &: \sigma p - \tau = 0 , \\
\delta\alpha &: \phi_x - v = 0 ,
\end{aligned}$$

T. J. Burchell, T. J. Bridges, *Transverse Instability of Solitary Waves*, Mathematical Physics Studies, https://doi.org/10.1007/978-981-95-2046-6

$$\begin{aligned} \delta\beta &: v_x - u = 0\,, \\ \delta\gamma &: u_x - w = 0\,, \\ \delta\tau &: v_y - p = 0\,. \end{aligned}$$

Two of the Lagrange multipliers can be eliminated: $\gamma = -w$, and $\tau = \sigma p$, leading to

$$L = \frac{1}{2}uv_t + F(u) + \frac{1}{2}w^2 - \frac{1}{2}\sigma p^2 + \alpha\phi_x - \alpha v + \beta v_x - \beta u - wu_x + \sigma p v_y\,.$$

The phase space is now seven–dimensional. The constraint

$$u_y = p_x$$

is implicit in that variations are required to satisfy it. However, it can be made explicit, with the added advantage of introducing a new dependent variable, thereby making the phase space eight–dimensional. Adding the above constraint with Lagrange multiplier κ gives

$$L = \frac{1}{2}uv_t + F(u) + \frac{1}{2}w^2 - \frac{1}{2}\sigma p^2 + \alpha\phi_x - \alpha v + \beta v_x - \beta u - wu_x + \sigma p v_y + \kappa(u_y - p_x)\,.$$

This Lagrangian now depends on $\phi, u, v, w, p, \alpha, \beta, \gamma$, and κ. Taking variations with respect to each coordinate in the cube C with fixed endpoint conditions gives

$$\begin{aligned} \delta\phi &: -\alpha_x = 0\,, \\ \delta v &: -\tfrac{1}{2}u_t - \alpha - \beta_x - \sigma p_y = 0\,, \\ \delta u &: \tfrac{1}{2}v_t + f(u) - \beta - \kappa_y = 0\,, \\ \delta w &: w - u_x = 0\,, \\ \delta p &: -\sigma p + \kappa_x = 0\,, \\ \delta\alpha &: \phi_x - v = 0\,, \\ \delta\beta &: v_x - u = 0\,, \\ \delta\kappa &: u_y - p_x = 0\,. \end{aligned}$$

Now re-label coordinates to make for a more tidy formulation; let

$$\phi := q_1, \quad v = q_2, \quad w = q_3, \quad \kappa = q_4, \quad \alpha = p_1, \quad \beta = p_2, \quad u = p_3, \quad p = p_4.$$

This way, the phase space is $\mathbb{R}^8$ with coordinates $(q_1, q_2, q_3, q_3, p_1, p_2, p_3, p_4)$. The Lagrangian density is now

$$L = \tfrac{1}{2}p_3(q_2)_t + F(p_3) + \tfrac{1}{2}(q_3)^2 - \tfrac{1}{2}\sigma(p_4)^2 + p_1(q_1)_x - p_1q_2 + p_2(q_2)_x$$
$$-p_2p_3 - q_3(p_3)_x + \sigma p(q_2)_y - q_4(p_4)_x + q_4(p_3)_y\,.$$

Partition terms according to derivatives:

$$L = \underbrace{\frac{1}{2}p_3(q_2)_t}_{t\text{ derivatives}} + \underbrace{p_1(q_1)_x + p_2(q_2)_x - q_3(p_3)_x - q_4(p_4)_x}_{x\text{ derivatives}} + \underbrace{\sigma p_4(q_2)_y + q_4(p_3)_y}_{y\text{ derivatives}}$$
$$+ \underbrace{F(p_3) + \frac{1}{2}(q_3)^2 - \frac{1}{2}\sigma(p_4)^2 - p_1q_2 - p_2p_3}_{\text{algebraic terms}}\,.$$

Taking variations with respect to $Z := (\mathbf{q}, \mathbf{p}) \in \mathbb{R}^8$,

$$\begin{aligned}
\delta q_1 &: -(p_1)_x = 0\,,\\
\delta q_2 &: -\tfrac{1}{2}(p_3)_t - (p_2)_x - \sigma(p_4)_y - p_1 = 0\,,\\
\delta q_3 &: -(p_3)_x + q_3 = 0\,,\\
\delta q_4 &: -(p_4)_x + (p_3)_y = 0\,,\\
\delta p_1 &: (q_1)_x - q_2 = 0\,,\\
\delta p_2 &: (q_2)_x - p_3 = 0\,,\\
\delta p_3 &: \tfrac{1}{2}(q_2)_t + (q_3)_x - (q_4)_y + f(p_3) - p_2 = 0\,,\\
\delta p_4 &: +(q_4)_x + \sigma(q_2)_y - \sigma p_4 = 0\,.
\end{aligned}$$

This system of PDEs is of the form

$$\mathbf{M}Z_t + \mathbf{K}Z_x + \mathbf{N}Z_y = \nabla S(Z)\,, \quad Z \in \mathbb{R}^8\,,$$

with $S(Z)$, and the structure matrices $\mathbf{M}$, $\mathbf{K}$, and $\mathbf{N}$ given in Sect. 3.8.

Written out, the equations are

$$
\begin{aligned}
(q_1)_x &= q_2\,,\\
(q_2)_x &= p_3\,,\\
(p_3)_x &= q_3\,,\\
(p_1)_x &= 0\,,\\
(p_3)_y &= (p_4)_x\,,
\end{aligned}
$$

along with the definitions

$$
\begin{aligned}
p_1 &= -\frac{1}{2}(p_3)_t - (p_2)_x - \sigma(p_4)_y,\\
p_2 &= \frac{1}{2}(q_2)_t + (q_3)_x - (q_4)_y + f(p_3),\\
p_4 &= (q_2)_y + \frac{1}{\sigma}(q_4)_x\,.
\end{aligned}
$$

This formulation is equivalent to that of Liu and Qin [104], derived using a jet-bundle formulation, but is more compact as the phase space dimension is eight as opposed to ten in [104]. With the reduction in phase space, the structure operator **K** is non-degenerate.

Appendix B
Spectral Problem with sech^2 Potential

This appendix establishes the basic properties of the elementary spectral problem

$$\phi_{xx} + 12\text{sech}^2(x)\,\phi = \Lambda\phi\,, \tag{B.1}$$

that appears in the examples in the introduction, and in Chap. 9. The solutions of this ODE can be determined explicitly using an Evans-function approach. The calculations are elementary and the result is just summarised. The eigenvalues of (B.1) are $\Lambda = 1, 4, 9$, and for all $\Lambda \notin \{1, 4, 9\}$ the two functions

$$\phi^{\pm}(x, \Lambda) = \mathrm{e}^{\pm\sqrt{\Lambda}x}\left(\pm a_0 + a_1\tanh(x) \pm a_2\tanh^2(x) + \tanh^3(x)\right)$$

are linearly independent, where

$$a_0 = \frac{\sqrt{\Lambda}}{15}(4 - \Lambda)\,, \quad a_1 = \frac{1}{5}(2\Lambda - 3)\,, \quad a_2 = -\sqrt{\Lambda}\,.$$

That $\Lambda = 1, 4, 9$ are the only eigenvalues, and that $\phi^{\pm}$ are linearly independent is verified by computing the Wronskian

$$W(x, \Lambda) = \det\begin{bmatrix} \phi^+ & \phi^- \\ \phi_x^+ & \phi_x^- \end{bmatrix}.$$

It is easily verified that $W_x = 0$ and so $W(x, \Lambda)$ is independent of x. Evaluate at $x = 0$,

$$W(0, \Lambda) = \det\begin{bmatrix} a_0 & -a_0 \\ a_1 + a_0\sqrt{\Lambda}a_1 & a_0\sqrt{\Lambda} \end{bmatrix} = 2a_0\big(a_1 + a_0\sqrt{\Lambda}\big)\,.$$

T. J. Burchell, T. J. Bridges, *Transverse Instability of Solitary Waves*, Mathematical Physics Studies, https://doi.org/10.1007/978-981-95-2046-6

Substituting for a_0 and a_1,

$$W(0, \Lambda) = \frac{2}{225}\sqrt{\Lambda}(\Lambda - 1)(\Lambda - 4)(\Lambda - 9)\,.$$

Hence $\phi^{\pm}$ are linearly independent for all $\Lambda \in \mathbb{C}^{+} \setminus \{1, 4, 9\}$.

At the selected values, $\{1, 4, 9\}$, the eigenfunctions are

$$\begin{aligned}
\phi(x, 1) &= \operatorname{sech}(x)\big(4 - 5\operatorname{sech}^2(x)\big)\,, \quad \text{when } \Lambda = 1\,,\\
\phi(x, 4) &= \tanh(x)\operatorname{sech}^2(x)\,, \quad \text{when } \Lambda = 4\,,\\
\phi(x, 9) &= \operatorname{sech}^3(x)\,, \quad \text{when } \Lambda = 9\,,
\end{aligned}$$

modulo an arbitrary multiplicative constant.

Appendix C
Choosing the Coupling Terms for the 2+1 Example

To determine the best choice of coupling terms for Eqs. (9.3)–(9.6) to ensure they have exact solutions and will decouple with a simple transformation, start by considering a general form of the initial equations

$$\phi_{tt} - \phi_{xx} - \phi_{yy} + 4\phi - 6\phi^2 + pQ_\phi = 0\,, \tag{C.1}$$

$$(v_1)_{tt} - (v_1)_{xx} - (v_1)_{yy} + 4v_1 - 6{\beta_1}^{-1}{v_1}^2 + pQ_{v_1} = 0\,, \tag{C.2}$$

$$(v_2)_{tt} - (v_2)_{xx} - (v_2)_{yy} + 4v_2 - 6{\beta_2}^{-1}{v_2}^2 - pQ_{v_2} = 0\,, \tag{C.3}$$

$$(v_3)_{tt} - (v_3)_{xx} - (v_3)_{yy} + 4v_3 - 6{\beta_3}^{-1}{v_3}^2 - pQ_{v_3} = 0\,, \tag{C.4}$$

where $p \in \mathbb{R}$ is a parameter, $\beta_j \in \mathbb{R}$ are non-zero constants to be determined and $Q(\phi, v_1, v_2, v_3)$ is a quadratic function of each of its inputs with no linear or constant terms. Because of this, each derivative of Q will be a linear combination of ϕ, v_1, v_2, v_3, while $\mathbf{Q} := D^2 Q$ will be a constant symmetric 4×4 matrix. Now, if

$$Q_\phi = Q_{v_1} = Q_{v_2} = Q_{v_3} = 0 \quad \text{when} \quad v_j = \beta_j \phi\,, \quad j = 1, 2, 3\,, \tag{C.5}$$

this will lead to a simple basic state $\phi = \hat{\phi},\ v_j = \beta_j \hat{\phi}$, where $\hat{\phi}$ satisfies

$$\hat{\phi}_{tt} - \hat{\phi}_{xx} - \hat{\phi}_{yy} + 4\hat{\phi} - 6\hat{\phi}^2 = 0. \tag{C.6}$$

T. J. Burchell, T. J. Bridges, *Transverse Instability of Solitary Waves*, Mathematical Physics Studies, https://doi.org/10.1007/978-981-95-2046-6

(Note that this equation will be solved by $\hat{\phi}$ as given in (9.12).) It is easy to check that the condition in (C.5) is equivalent to

$$\mathbf{Q}\begin{pmatrix}1\\ \beta_1\\ \beta_2\\ \beta_3\end{pmatrix} = \mathbf{0}\,, \tag{C.7}$$

i.e. $(1, \beta_1, \beta_2, \beta_3)$ is an eigenvector of $\mathbf{Q}$ corresponding to a zero eigenvalue.

C.1 Applying to the Linearisation

Linearising about the basic state leads to this system (the general form of (9.31)–(9.34)):

$$L\begin{pmatrix}\tilde{\phi}\\ \tilde{v}_1\\ \tilde{v}_2\\ \tilde{v}_3\end{pmatrix} = p\tilde{\mathbf{R}}\mathbf{Q}\begin{pmatrix}\tilde{\phi}\\ \tilde{v}_1\\ \tilde{v}_2\\ \tilde{v}_3\end{pmatrix}, \tag{C.8}$$

where L is defined the same way as in (9.31)–(9.34) and

$$\tilde{\mathbf{R}} = \mathrm{diag}(1, 1, -1, -1).$$

The first derivative terms can be scaled out by taking the transformation

$$\begin{pmatrix}\tilde{\phi}\\ \tilde{v}_1\\ \tilde{v}_2\\ \tilde{v}_3\end{pmatrix} = e^{\alpha^2\Upsilon\xi}\begin{pmatrix}\Psi_1\\ \Psi_2\\ \Psi_3\\ \Psi_4\end{pmatrix}, \qquad \Upsilon = c\lambda - \mathrm{i}\ell\nu\,,$$

leaving us with

$$\left(\alpha^{-2}\partial_{\xi\xi} + \widehat{\varphi} - \widehat{\Upsilon}\right)\begin{pmatrix}\Psi_1\\ \Psi_2\\ \Psi_3\\ \Psi_4\end{pmatrix} = p\tilde{\mathbf{R}}\mathbf{Q}\begin{pmatrix}\Psi_1\\ \Psi_2\\ \Psi_3\\ \Psi_4\end{pmatrix}, \qquad \widehat{\Upsilon} = \alpha^2\Upsilon^2 + \lambda^2 + \nu^2. \tag{C.9}$$

Now in order to decouple this system we need to diagonalise the matrix on the RHS. We can see from (C.7) that $(1, \beta_1, \beta_2, \beta_3)$ is one eigenvector of $\tilde{\mathbf{R}}\mathbf{Q}$ (with corresponding eigenvalue 0). For simplicity, suppose that the other eigenvectors are

$(\beta_1, 1, \beta_3, \beta_2)$, $(\beta_2, \beta_3, 1, \beta_1)$ and $(\beta_3, \beta_2, \beta_1, 1)$ with corresponding eigenvalues ρ_1, ρ_2, ρ_3. This allows us to write

$$\mathbf{Q} = \tilde{\mathbf{R}}\mathbf{PDP}^{-1}, \tag{C.10}$$

where

$$\mathbf{P} = \begin{pmatrix} 1 & \beta_1 & \beta_2 & \beta_3 \\ \beta_1 & 1 & \beta_3 & \beta_2 \\ \beta_2 & \beta_3 & 1 & \beta_1 \\ \beta_3 & \beta_2 & \beta_1 & 1 \end{pmatrix}, \qquad \mathbf{D} = \text{diag}(0, \rho_1, \rho_2, \rho_3).$$

Applying the transformation

$$\begin{pmatrix} \Psi_1 \\ \Psi_2 \\ \Psi_3 \\ \Psi_4 \end{pmatrix} = \mathbf{P} \begin{pmatrix} \psi_1 \\ \psi_2 \\ \psi_3 \\ \psi_4 \end{pmatrix}$$

to (C.9) then results in

$$\left(\alpha^{-2}\partial_{\xi\xi} + \widehat{\varphi} - \widehat{\Upsilon}\right) \begin{pmatrix} \psi_1 \\ \psi_2 \\ \psi_3 \\ \psi_4 \end{pmatrix} = p \begin{pmatrix} 0 \\ \rho_1\psi_2 \\ \rho_2\psi_3 \\ \rho_3\psi_4 \end{pmatrix}, \tag{C.11}$$

which is a general form of (9.36)–(9.39). This allows us to write the full transformation from the coupled system (C.8) to the decoupled one (C.11) as

$$\begin{pmatrix} \tilde{\phi} \\ \tilde{v}_1 \\ \tilde{v}_2 \\ \tilde{v}_3 \end{pmatrix} = e^{\alpha^2\Upsilon\xi}\mathbf{P} \begin{pmatrix} \psi_1 \\ \psi_2 \\ \psi_3 \\ \psi_4 \end{pmatrix}, \tag{C.12}$$

providing the origin of the transformation in (9.35) (with the values of $\beta_1, \beta_2, \beta_3$ as specified in the next section).

It can be shown that $\hat{\phi}_\xi$ is a solution for ψ_1 when $\lambda, \nu = 0$ so the equation of the first component will generate the basic solution $\widehat{Z}_\xi$. Therefore, the solutions of the equations for the other 3 components (at $\lambda, \nu = 0$) will generate $\mathbf{a}^\pm$, $\mathbf{b}^\pm$, $\mathbf{c}^\pm$ and by considering how these solutions behave we can deduce that $\rho_1, \rho_2, \rho_3 > 0$. We can also see that ρ_1, ρ_2, ρ_3 need to be distinct, otherwise the solutions will not be distinct which would mean that $\Pi \equiv 0$. For the example and w.l.o.g. we can assume that $\rho_1 < \rho_2 < \rho_3$ (other possibilities can be created from this by relabelling). This will correspond to the solutions for ψ_2, ψ_3 and ψ_4 respectively generating $\mathbf{a}^{\pm}$, $\mathbf{b}^{\pm}$ and $\mathbf{c}^{\pm}$.

C.2 Placing Constraints on the β's

Now in order for this to work it is necessary to place two conditions on the possible values of $\beta_1, \beta_2, \beta_3$:

$$(a) \qquad \det \mathbf{P} \neq 0,$$

$$(b) \qquad \tilde{\mathbf{R}}\mathbf{P}\mathbf{D}\mathbf{P}^{-1} \text{ is symmetric.}$$

It can be shown that (b) will hold if and only if $\rho_1 = 0,\ \rho_2 = \rho_3$ OR $\beta_1 = \beta_2\beta_3$. Since we want the ρ's to be distinct and non-zero we will take the latter condition. Substituting this into $\det \mathbf{P}$ results in

$$\det \mathbf{P} = ({\beta_2}^2 - 1)^2({\beta_3}^2 - 1)^2,$$

which is non-zero for $\beta_2, \beta_3 \neq \pm 1$. It can be checked that $\mathbf{P}^{-1}$ in this case will be

$$\mathbf{P}^{-1} = \frac{1}{({\beta_2}^2 - 1)({\beta_3}^2 - 1)} \begin{pmatrix} 1 & \beta_2\beta_3 & -\beta_2 & -\beta_3 \\ \beta_2\beta_3 & 1 & -\beta_3 & -\beta_2 \\ -\beta_2 & -\beta_3 & 1 & \beta_2\beta_3 \\ -\beta_3 & -\beta_2 & \beta_2\beta_3 & 1 \end{pmatrix}.$$

To satisfy these conditions, let us choose the following values for the example:

$$\beta_1 = 3, \qquad \beta_2 = 2, \qquad \beta_3 = \frac{3}{2},$$

$$\rho_1 = 5, \qquad \rho_2 = 10, \qquad \rho_3 = 15\,,$$

which after substituting into (C.10) results in

$$D^2Q := \mathbf{Q} = \frac{1}{3} \begin{pmatrix} -23 & -48 & 52 & 42 \\ -48 & -62 & 78 & 52 \\ 52 & 78 & -107 & -48 \\ 42 & 52 & -48 & -68 \end{pmatrix}. \tag{C.13}$$

Therefore,

$$
\begin{aligned}
Q_\phi &= -\frac{23}{3}\phi - 16v_1 + \frac{52}{3}v_2 + 14v_3\,, \\
Q_{v_1} &= -16\phi - \frac{62}{3}v_1 + 26v_2 + \frac{52}{3}v_3\,, \\
Q_{v_2} &= \frac{52}{3}\phi + 26v_1 - \frac{107}{3}v_2 - 16v_3\,, \\
Q_{v_3} &= 14\phi + \frac{52}{3}v_1 - 16v_2 - \frac{68}{3}v_3\,,
\end{aligned}
\tag{C.14}
$$

which produces the coupling terms seen in (9.3)–(9.6).

Appendix D
Parity of the Maslov Index of Solitary Waves

In this appendix we give a sketch of the proof, following [38–40], of the connection between transversality and parity of the Maslov index given in the formula (1.39). We use a construction of the Maslov index of solitary waves based on Souriau's definition (cf. Souriau [133] and Chapter 3 of de Gosson [60]). It is formulated on the "Maslov bundle", which is also the universal cover of the Lagrangian Grassmannian (cf. §3.2.2 of [60]). A Lagrangian plane $\mathbf{U} \in \mathrm{Lag}(n)$, in a $2n$-dimensional real vector space V with standard symplectic form, can be represented by a $2n \times n$ matrix of rank n,

$$\mathbf{U} = \begin{bmatrix} \mathbf{U}_1 \\ \mathbf{U}_2 \end{bmatrix}, \quad \text{with} \quad \mathbf{U}_1^T \mathbf{U}_2 = \mathbf{U}_2^T \mathbf{U}_1 \,.$$

The columns of $\mathbf{U}$ form an orthonormal basis, for a Lagrangian plane, if and only if the matrix $\mathbf{u} := \mathbf{U}_1 + \mathrm{i}\mathbf{U}_2 \in U(n)$, where $U(n)$ is the set of $n \times n$ unitary matrices. The Maslov bundle has total space $\widetilde{U(n)}$ (the universal cover of the Lagrangian Grassmannian), base manifold $\mathrm{Lag}(n)$, and projection π,

$$\pi : \widetilde{U(n)} \to \mathrm{Lag}(n)\,, \quad (\mathbf{U}, \kappa) \mapsto \text{the space spanned by } \mathbf{U}\,,$$

with

$$\widetilde{U(n)} = \left\{ (\mathbf{U}, \kappa) \;\middle|\; \mathbf{U} = \begin{bmatrix} \mathbf{U}_1 \\ \mathbf{U}_2 \end{bmatrix}, \; \mathbf{u} = \mathbf{U}_1 + \mathrm{i}\mathbf{U}_1 \in U(n)\,, \; \mathrm{e}^{-\mathrm{i}\kappa/2} = \det(\mathbf{u}) \right\}. \tag{D.1}$$

Denote by $W(n, \mathbb{C})$ the set of all *symmetric* unitary matrices

$$W(n, \mathbb{C}) := \left\{ \, \mathbf{w} \in U(n) \; : \; \mathbf{w} = \mathbf{w}^T \, \right\}. \tag{D.2}$$

T. J. Burchell, T. J. Bridges, *Transverse Instability of Solitary Waves*, Mathematical Physics Studies, https://doi.org/10.1007/978-981-95-2046-6

The product of two symmetric matrices is not necessarily symmetric, and so $W(n, \mathbb{C})$ is not a group, although $W(n, \mathbb{C})$ is closed under the operation of taking square roots (cf. Lemma 3.1 in [60]). Moreover, symmetry and the condition $\mathbf{w}^H\mathbf{w} = \mathbf{I}$ give that

$$\mathbf{w}_1\mathbf{w}_2 = \mathbf{w}_2\mathbf{w}_1 \quad \text{for} \quad \mathbf{w} = \mathbf{w}_1 + \mathrm{i}\mathbf{w}_2 \in W(n, \mathbb{C}).$$

It follows that the real symmetric matrices $\mathbf{w}_1$ and $\mathbf{w}_2$ can be diagonalised simultaneously: there exists an orthogonal matrix $\mathbf{R} \in O(n, \mathbb{R})$, such that $\mathbf{R}^{-1}\mathbf{w}_1\mathbf{R}$ and $\mathbf{R}^{-1}\mathbf{w}_2\mathbf{R}$ are diagonal.

We will also need the *Souriau mapping*

$$\psi : \mathrm{Lag}(n) \to W(n, \mathbb{C}), \quad \psi(\mathbf{u}) = \mathbf{u}\mathbf{u}^T. \tag{D.3}$$

The Souriau mapping has the properties

$$\psi^H = \overline{\psi}, \quad \psi^T = \psi, \quad \overline{\psi}\psi = \mathbf{I}, \tag{D.4}$$

and

$$\psi(\mathbf{v}\mathbf{u}) = \mathbf{v}\psi(\mathbf{u})\mathbf{v}^T, \quad \text{for any } \mathbf{v} \in U(n).$$

We will use the Souriau mapping for pairs of Lagrangian planes to construct the Maslov index. Let $\mathbf{U}$ and $\mathbf{V}$ be Lagrangian planes, with respective unitary representations $\mathbf{u}$ and $\mathbf{v}$. Then evaluate $\overline{\mathbf{u}}^{-1}\overline{\mathbf{v}}$ on the Souriau mapping

$$\begin{aligned}\Psi(\mathbf{U}, \mathbf{V}) &:= \psi(\overline{\mathbf{u}}^{-1}\overline{\mathbf{v}}) \\ &= \overline{\mathbf{u}}^{-1}\overline{\mathbf{v}}\mathbf{v}^{-1}\mathbf{u} \\ &= (\mathbf{U}_1 - \mathrm{i}\mathbf{U}_2)^{-1}(\mathbf{V}_1 - \mathrm{i}\mathbf{V}_2)(\mathbf{V}_1 + \mathrm{i}\mathbf{V}_2)^{-1}(\mathbf{U}_1 + \mathrm{i}\mathbf{U}_2).\end{aligned} \tag{D.5}$$

As the image of Ψ is in $W(n, \mathbb{C})$, the eigenvalues lie on the unit circle and can be expressed in the form:

$$\sigma\big(\Psi(\mathbf{U}, \mathbf{V})\big) = \big\{\mathrm{e}^{\mathrm{i}\alpha_1(\mathbf{U},\mathbf{V})}, \ldots, \mathrm{e}^{\mathrm{i}\alpha_n(\mathbf{U},\mathbf{V})}\big\}, \quad 0 \le \alpha_1 \le \ldots \le \alpha_n < 2\pi.$$

To lighten the notation, we will drop the arguments on $\alpha_1, \ldots, \alpha_n$ and their dependence will be clear from the context.

Proposition D.1 *There exists $n \times n$ orthogonal matrices $\mathbf{R}$ and $\mathbf{S}$ such that*

$$\mathbf{R}^{-1}\Psi(\mathbf{U}, \mathbf{V})\mathbf{R} = \mathbf{D}^2, \quad (\mathbf{U}_1\mathbf{R} + \mathrm{i}\mathbf{U}_2\mathbf{R})\overline{\mathbf{D}} = (\mathbf{V}_1\mathbf{S} + \mathrm{i}\mathbf{V}_2\mathbf{S}), \tag{D.6}$$

with

$$\mathbf{D} := \begin{bmatrix} e^{\frac{1}{2}\mathrm{i}\alpha_1} & & 0 \\ & \ddots & \\ 0 & & e^{\frac{1}{2}\mathrm{i}\alpha_n} \end{bmatrix}.$$

Proof Since Ψ is unitary and symmetric, the real and imaginary parts, $\Re\Psi$ and $\Im\Psi$ respectively, are simultaneously diagonalisable: there exists an $\mathbf{R} \in O(n, \mathbb{R})$ such that $\mathbf{R}^{-1}\Re\psi\mathbf{R}$ and $\mathbf{R}^{-1}\Im\psi\mathbf{R}$ are diagonal. By choosing an appropriate permutation of columns, $\mathbf{R}$ can be chosen such that

$$\mathbf{R}^{-1}\Psi(\mathbf{U}, \mathbf{V})\mathbf{R} = \mathbf{D}^2 = \mathbf{D}\overline{\mathbf{D}}^{-1} = \mathbf{D}\mathbf{D}^T.$$

Define

$$\mathbf{S} = (\mathbf{V}_1 + \mathrm{i}\mathbf{V}_2)^{-1}(\mathbf{U}_1 + \mathrm{i}\mathbf{U}_2)\mathbf{R}\overline{\mathbf{D}}.$$

Then using the definition of Ψ in (D.5), this is equivalent to

$$\begin{aligned} \mathbf{S} &= (\mathbf{V}_1 + \mathrm{i}\mathbf{V}_2)^{-1}(\mathbf{U}_1 + \mathrm{i}\mathbf{U}_2)\mathbf{R}\overline{\mathbf{D}} \\ &= (\mathbf{V}_1 - \mathrm{i}\mathbf{V}_2)^{-1}(\mathbf{U}_1 - \mathrm{i}\mathbf{U}_2)\Psi(\mathbf{U}, \mathbf{V})\mathbf{R}\overline{\mathbf{D}} \\ &= (\mathbf{V}_1 - \mathrm{i}\mathbf{V}_2)^{-1}(\mathbf{U}_1 - \mathrm{i}\mathbf{U}_2)\mathbf{R}\mathbf{D}^2\overline{\mathbf{D}} \\ &= (\mathbf{V}_1 - \mathrm{i}\mathbf{V}_2)^{-1}(\mathbf{U}_1 - \mathrm{i}\mathbf{U}_2)\mathbf{R}\mathbf{D}, \end{aligned}$$

and so

$$\begin{aligned} \overline{\mathbf{S}} &= \overline{(\mathbf{V}_1 - \mathrm{i}\mathbf{V}_2)^{-1}(\mathbf{U}_1 - \mathrm{i}\mathbf{U}_2)\mathbf{R}\mathbf{D}} \\ &= (\mathbf{V}_1 + \mathrm{i}\mathbf{V}_2)^{-1}(\mathbf{U}_1 + \mathrm{i}\mathbf{U}_2)\mathbf{R}\overline{\mathbf{D}} \\ &= \mathbf{S}. \end{aligned}$$

Similarly,

$$\begin{aligned} \mathbf{S}^T &= \overline{\mathbf{D}}^T\mathbf{R}^T(\mathbf{U}_1 + \mathrm{i}\mathbf{U}_2)^T(\mathbf{V}_1 + \mathrm{i}\mathbf{V}_2)^{-T} \\ &= \overline{\mathbf{D}}^T\mathbf{R}^{-1}(\mathbf{U}_1 - \mathrm{i}\mathbf{U}_2)^{-1}(\mathbf{V}_1 - \mathrm{i}\mathbf{V}_2) \\ &= \overline{\mathbf{D}}^T\mathbf{R}^{-1}\Psi(\mathbf{U}_1 + \mathrm{i}\mathbf{U}_2)^{-1}(\mathbf{V}_1 + \mathrm{i}\mathbf{V}_2) \\ &= \overline{\mathbf{D}}^T\mathbf{D}^2\mathbf{R}^{-1}(\mathbf{U}_1 + \mathrm{i}\mathbf{U}_2)^{-1}(\mathbf{V}_1 + \mathrm{i}\mathbf{V}_2) \\ &= \mathbf{D}\mathbf{R}^{-1}(\mathbf{U}_1 + \mathrm{i}\mathbf{U}_2)^{-1}(\mathbf{V}_1 + \mathrm{i}\mathbf{V}_2) \\ &= \mathbf{S}^{-1}. \end{aligned}$$

These two results prove that $\mathbf{S} \in O(n, \mathbb{R})$, and so

$$(\mathbf{U}_1\mathbf{R} + \mathrm{i}\mathbf{U}_2\mathbf{R})\overline{\mathbf{D}} = (\mathbf{V}_1\mathbf{S} + \mathrm{i}\mathbf{V}_2\mathbf{S}),$$

proving the second equation in (D.6). ∎

Proposition D.2 *Let* $\mathbf{r}_1, \ldots, \mathbf{r}_n$ *be the columns of the* $\mathbb{R}^{2n\times n}$ *matrix* **UR**, *and define* $\mathbf{r}_{n+i} = \mathbf{J}\mathbf{r}_i$. *Let* $\mathbf{s}_1, \ldots, \mathbf{s}_n$ *be the columns of the* $\mathbb{R}^{2n\times n}$ *matrix* **VS**. *Then:*

$$\mathbf{s}_i = \cos\left(\frac{\alpha_i}{2}\right)\mathbf{r}_i - \sin\left(\frac{\alpha_i}{2}\right)\mathbf{r}_{n+i}$$

Let d *be the multiplicity of* 1 *as an eigenvalue of* Ψ. *Then the intersection of the spaces spanned by* **U** *and* **V** *(which are the same as the spaces spanned by* **UR** *and* **VS**, *but not necessarily with the same orientation) is the space spanned by* $\mathbf{r}_1, \ldots, \mathbf{r}_d$.

Proof From (D.6) we have that

$$(\mathbf{U}_1 + \mathrm{i}\mathbf{U}_2)\mathbf{R}\overline{\mathbf{D}} = \big(\mathbf{V}_1 + \mathrm{i}\mathbf{V}_2\big)\mathbf{S}.$$

Writing this in real coordinates gives

$$\begin{pmatrix}\mathbf{U}_1\\ \mathbf{U}_2\end{pmatrix}\mathbf{R}(\Re\overline{\mathbf{D}}) + \begin{bmatrix}\mathbf{0} & -\mathbf{I}\\ \mathbf{I} & \mathbf{0}\end{bmatrix}\begin{pmatrix}\mathbf{U}_1\\ \mathbf{U}_2\end{pmatrix}\mathbf{R}(\Im\overline{\mathbf{D}}) = \begin{pmatrix}\mathbf{V}_1\\ \mathbf{V}_2\end{pmatrix}\mathbf{S}.$$

Using the definitions of $\mathbf{r}_j$, $j = 1, \ldots, 2n$,

$$\Big[\mathbf{r}_1\Big|\cdots\Big|\mathbf{r}_n\Big](\Re\overline{\mathbf{D}}) + \Big[\mathbf{r}_{n+1}\Big|\cdots\Big|\mathbf{r}_{2n}\Big](\Im\overline{\mathbf{D}}) = \Big[\mathbf{s}_1\Big|\cdots\Big|\mathbf{s}_n\Big].$$

It follows that

$$\mathbf{s}_i = \cos\left(\frac{\alpha_i}{2}\right)\mathbf{r}_i - \sin\left(\frac{\alpha_i}{2}\right)\mathbf{r}_{n+i}, \quad i = 1, \ldots, n.$$

Hence,

$$\begin{aligned}\mathbf{\Omega}(\mathbf{r}_i, \mathbf{s}_j) &= \left\langle\mathbf{J}\mathbf{r}_i, \cos\left(\frac{\alpha_j}{2}\right)\mathbf{r}_j - \sin\left(\frac{\alpha_j}{2}\right)\mathbf{J}\mathbf{r}_j\right\rangle = -\sin\left(\frac{\alpha_j}{2}\right)\langle\mathbf{r}_i, \mathbf{r}_j\rangle\\ &= -\sin\left(\frac{\alpha_i}{2}\right)\delta_{ij}.\end{aligned}$$

We have that

$$\mathrm{span}\{\mathbf{r}_1, \ldots, \mathbf{r}_d\} = \mathrm{span}\{\mathbf{s}_1, \ldots, \mathbf{s}_d\},$$

and

$$\mathbf{r}_1 \wedge \ldots \wedge \mathbf{r}_n \wedge \mathbf{s}_{d+1} \wedge \ldots \wedge \mathbf{s}_n$$
$$= (-1)^{n-d} \sin(\alpha_{d+1}) \cdot \ldots \cdot \sin(\alpha_n) \mathbf{r}_1 \wedge \ldots \wedge \mathbf{r}_n \wedge \mathbf{r}_{n+d+1} \wedge \ldots \wedge \mathbf{r}_{2n} \neq 0 .$$

We conclude that $\pi(\mathbf{U}, \kappa) \cap \pi(\mathbf{V}, \tau)$ is the space spanned by $\mathbf{r}_1, \ldots, \mathbf{r}_d$. ■

According to Souriau's formula (cf. pages 126–128 of [133]), the Maslov index of this pair of elements is defined in the following way.

Definition 9 Let $(\mathbf{U}, \kappa)$ and $(\mathbf{V}, \tau)$ be in $\widetilde{U(n)}$. The Maslov index of this pair of elements is defined by:

$$\mathsf{Maslov}\big((\mathbf{U}, \kappa), (\mathbf{V}, \tau)\big) = \frac{\tau - \kappa}{2\pi} - \frac{\alpha_1 + \cdots + \alpha_n}{2\pi} + \frac{1}{2} d , \tag{D.7}$$

where

$$d := \dim\big(\pi((\mathbf{U}, \kappa)) \cap \pi((\mathbf{V}, \tau))\big) .$$

Remark It is essential that $\alpha_1, \ldots, \alpha_n \in [0, 2\pi)$, or there would be several possible values of the Maslov index. It is clear that Maslov is an integer when d is even and an half-integer when d is odd.

The definition for the Maslov index of a homoclinic orbit, based on Souriau's definition above, is as follows. Let $\widehat{\mathbf{u}}$ be an homoclinic orbit. Let $(\mathbf{U}^+, \kappa^+)$ and $(\mathbf{U}^-, \kappa^-)$ in $\widetilde{U(n)}$ be such that:

$$\begin{cases} \operatorname{span}\Big(\mathbf{U}^+(x)\Big) = \operatorname{span}\{\widehat{\mathbf{u}}_x(x), \mathbf{a}_1^+(x), \ldots, \mathbf{a}_{n-1}^+(x)\} \\ \operatorname{span}\Big(\mathbf{U}^-(x)\Big) = \operatorname{span}\{\widehat{\mathbf{u}}_x(x), \mathbf{a}_1^-(x), \ldots, \mathbf{a}_{n-1}^-(x)\} \\ \kappa^+, \kappa^- \text{are continuous} . \end{cases}$$

Then the Maslov index of the homoclinic is

$$\mathsf{Maslov}^{hom}(\widehat{\mathbf{u}}) = \mathsf{Maslov}\big((\mathbf{U}^-(x), \kappa^-(x)), (\mathbf{U}^+(x), \kappa^+(x))\big) - \lim_{y \to +\infty} \mathsf{Maslov}\big((\mathbf{U}^-(-y), \kappa^-(-y)), (\mathbf{U}^+(y), \kappa^+(y))\big) , \tag{D.8}$$

and the definition is independent of x. In this definition, it is important that κ^+, κ^- are continuous. As a consequence, these functions are unique up to a shift by a multiple of 4π. If we shift κ^+ (resp. κ^-) by $4k\pi$, then the left of the minus sign is shifted by $2k$ (resp. $-2k$) and the right of the minus is shifted by $-2k$ (resp. $2k$). This guarantees that Maslov^{hom} is independent of the choice of κ^+ and κ^-.

Theorem D.3 *Suppose* $\widehat{\mathbf{u}}$ *is a transversely constructed homoclinic orbit with Maslov index* $\mathsf{Maslov}^{hom}(\widehat{\mathbf{u}})$. *Then* $d = 1$ *and*

$$(-1)^{\mathsf{Maslov}^{hom}+\frac{1}{2}} = \operatorname{sign}(\Pi)\,. \tag{D.9}$$

Remark 10 Since $d = 1$, the exponent $\mathsf{Maslov}^{hom} + \frac{1}{2}$ is an integer. Henceforth we will denote $\mathsf{Maslov}^{hom} + \frac{1}{2}$ by "Maslov". This difference between this Maslov index and the general definition (D.7) should be clear from the context.

The key point needed to prove the theorem is the relationship between the Maslov index and the intersection index for the tangent spaces of the stable and unstable subspaces.

Lemma D.4 *Let* $(\mathbf{U}, \kappa), (\mathbf{V}, \tau) \in \widetilde{U(n)}$ *such that* $d = \dim\big(\pi((\mathbf{U},\kappa)) \cap \pi((\mathbf{V},\tau))\big)$ *with* $d = 0, 1$, *and let* $\mathbf{U}^\wedge$ *and* $\mathbf{V}^\wedge$ *be the corresponding n-forms. Then*

$$\mathbf{O}_d(\mathbf{U}^\wedge, \mathbf{V}^\wedge) = (-1)^{m+\frac{d}{2}+n}\,, \quad m := \mathsf{Maslov}((\mathbf{U},\kappa),(\mathbf{V},\tau))\,.$$

Proof Let $\mathbf{R}, \mathbf{S}, \mathbf{D}, \mathbf{r}_i, \mathbf{s}_i$ be as defined in Propositions D.1 and D.2. Then,

$$\begin{aligned}
\mathbf{O}_d(\mathbf{U}\mathbf{R}^\wedge, \mathbf{V}\mathbf{S}^\wedge) &= \operatorname{sign}\det\begin{bmatrix} \boldsymbol{\Omega}(\mathbf{r}_{d+1}, \mathbf{s}_{d+1}) & \dots & \boldsymbol{\Omega}(\mathbf{r}_{d+1}, \mathbf{s}_n) \\ \vdots & & \vdots \\ \boldsymbol{\Omega}(\mathbf{r}_n, \mathbf{s}_{d+1}) & \dots & \boldsymbol{\Omega}(\mathbf{r}_n, \mathbf{s}_n) \end{bmatrix} \\
&= \operatorname{sign}\det\begin{bmatrix} -\sin(\frac{\alpha_{d+1}}{2}) & & 0 \\ & \ddots & \\ 0 & & -\sin(\frac{\alpha_n}{2}) \end{bmatrix} = (-1)^{n-d}\,.
\end{aligned}$$

As a consequence we have that:

$$\begin{aligned}
\mathbf{O}_d(\mathbf{U}^\wedge, \mathbf{V}^\wedge) &= \operatorname{sign}(\det \mathbf{R}\mathbf{S})\mathbf{O}_d(\mathbf{U}^\wedge, \mathbf{V}^\wedge) \\
&= \operatorname{sign}\det((\mathbf{U}_1 + \mathtt{i}\mathbf{U}_2)(\mathbf{V}_1 + \mathtt{i}\mathbf{V}_2)^{-1}\overline{\mathbf{D}})(-1)^{d+n} \\
&= e^{-\frac{1}{2}\mathtt{i}(\kappa-\tau)}(-1)^d e^{-\frac{1}{2}\mathtt{i}(\alpha_1+\cdots+\alpha_n)}(-1)^d = e^{\mathtt{i}\pi(m-\frac{1}{2}d)}(-1)^{d+n} \\
&= (-1)^{m+\frac{1}{2}d+n}\,.
\end{aligned}$$

and so, when $d = 1$ we have

$$\mathbf{O}_1(\mathbf{U}^\wedge, \mathbf{V}^\wedge) = (-1)^{m+\frac{1}{2}+n}\,.$$

∎

Proof (Of the Theorem) Use Lemma D.4 above to conclude the proof of Theorem D.3,

$$
\begin{aligned}
(-1)^{\mathsf{Maslov}^{hom}+\frac{1}{2}} = &\ \mathbf{O}_1\big((\widehat{\mathbf{u}}_x \wedge \mathbf{a}_1^-(x) \wedge \ldots \wedge \mathbf{a}_{n-1}^-(x)), (\widehat{\mathbf{u}}_x \wedge \mathbf{a}_1^+(x) \wedge \ldots \wedge \mathbf{a}_{n-1}^+(x))\big) \\
&\times \lim_{y\to+\infty} \mathbf{O}_0\big((\widehat{\mathbf{u}}_x \wedge \mathbf{a}_2^-(-y) \wedge \ldots \wedge \mathbf{a}_n^-(-y)), \\
&(\widehat{\mathbf{u}}_x(y) \wedge \mathbf{a}_2^+(y) \wedge \ldots \wedge \mathbf{a}_n^+(y))\big) \\
&\times(-1)^{\frac{1}{2}-(n+\frac{1}{2})-n} .
\end{aligned}
\tag{D.10}
$$

The last term of the product is $+1$. Hence the right-hand side is just the sign of Π. This completes the proof of the formula (D.9). ■

References

1. Abid, M., Kharif, C., Hsu, H.C., Chen, Y.Y.: Transverse instability of gravity-capillary solitary waves on deep water in the presence of constant vorticity. J. Fluid Mech. **871**, 1028–1043 (2019)
2. Akhmediev, N., Ankiewicz, A., Grimshaw, R.: Hamiltonian-versus-energy diagrams in soliton theory. Phys. Rev. E **59**, 6088–6096 (1999)
3. Akylas, T.R.: Envelope solitons with stationary crests. Phys. Fluids **5**, 789–791 (1993)
4. Alexander, J.W., Gardner, R., Jones, C.K.R.T.: A topological invariant arising in the stability analysis of traveling waves. J. Reine Angew. Math. **410**, 167–212 (1990)
5. Alexander, J.C., Pego, R.L., Sachs, R.L.: On the transverse instability of solitary waves in the Kadomtsev-Petviashvili equation. Phys. Lett. A **226**, 187–192 (1997)
6. Allen, M.A., Rowlands, G.: Stability of obliquely propagating plane solitons of the Zakharov-Kuznetsov equation. J. Plasma Phys. **53**, 63–73 (1995)
7. Barutello, V.L., Offin, D., Portaluri, A., Wu, L.: Sturm theory with applications in geometry and classical mechanics. Math. Z. **299**, 257–297 (2021)
8. Beck, M.: An introduction to recent applications of the Maslov index in dynamical systems and PDEs, Talk at the Joint Mathematics Meetings, Boston, https://math.bu.edu/people/mabeck/jmm-2023.pdf (2023)
9. Beck, M., Jaquette, J.: Validated spectral stability via conjugate points, SIAM J. Appl. Dyn. Sys. **21**, 366–404 (2022)
10. Beck, M., Cox, G., Jones, C., Latushkin, Y., McQuighan, K., Sukhtayev, A.: Instability of pulses in gradient reaction-diffusion systems: a symplectic approach. Philos. Trans. R. Soc. Lond. A **376**, 20170187 (2018)
11. Benzoni-Gavage, S.: Spectral transverse instability of solitary waves in Korteweg fluids. J. Math. Anal. Appl. **361**, 338–357 (2010)
12. Berkolaiko, G., Comech, A., Sukhtayev, A.: Vakhitov-Kolokolov and energy vanishing conditions for linear instability of solitary waves in models of classical self-interacting spinor fields. Nonlinearity **28**, 577–592 (2015)
13. Berryman, J.G.: Stability of solitary waves in shallow water. Phys. Fluids **19**, 771–777 (1976)
14. Bhavna, Kumar, A., Pandey, A.K.: Transverse spectral instability in generalized Kadomtsev–Petviashvili equation. Philos. Trans. R. Soc. Lond. A **478**, 20210693 (2022)
15. Bhavna, Pandey, A.K., Semenova, A.: Transverse spectral instabilities in rotation-modified Kadomtsev-Petviashvili equation and related models. Preprint. arXiv:2305.10700 (2023)
16. Bhavna, Pandey, A.K., Singh, S.: Transverse spectral instabilities in Konopelchenko-Dubrovsky equation. Stud. Appl. Math. **151**, 1053–1071 (2023)
17. Binz, E., Sniatycki, J., Fischer, H.: Geometry of Classical Fields. Dover Publications (1988)

T. J. Burchell, T. J. Bridges, *Transverse Instability of Solitary Waves*, Mathematical Physics Studies, https://doi.org/10.1007/978-981-95-2046-6

18. Blyuss, K.B.: Perturbed multi-symplectic systems: intersections of invariant manifolds and transverse instability. PhD Thesis, University of Surrey (2004)
19. Blyuss, K.B., Bridges, T.J., Derks, G.: Transverse instability and its long-term development for solitary waves of the 2+1-dimensional Boussinesq equation. Phys. Rev. E **67**, 056626 (2003)
20. Bona, J.L., Souganidis, P.E., Strauss, W.A.: Stability and instability of solitary waves of Korteweg-de Vries type. Proc. R. Soc. Lond. A **411**, 395–412 (1987)
21. Borluk, H., Bruell, G., Nilsson, D.: Traveling waves and transverse instability for the fractional Kadomtsev-Petviashvili equation. Stud. Appl. Math. **149**, 96–123 (2022)
22. Bose, A., Jones, C.K.R.T.: Stability of in-phase travelling wave solutions in a pair of coupled nerve fibers. Ind. Univ. Math. J. **44**, 189–220 (1995)
23. Bridges, T.J.: A new framework for studying the stability of genus-1 and genus-2 KP patterns. Euro. J. Mech. B/Fluids **18**, 493–500 (1999)
24. Bridges, T.J.: Universal geometric condition for the transverse instability of solitary waves. Phys. Rev. Lett. **84**, 2614–2618 (2000)
25. Bridges, T.J.: Transverse instability of solitary-wave states of the water-wave problem. J. Fluid Mech. **439**, 255–278 (2001)
26. Bridges, T.J.: On the susceptibility of bright nonlinear Schrödinger solitons to long-wave transverse instability. Proc. R. Soc. Lond. A **460**, 2505–2615 (2004)
27. Bridges, T.J.: Canonical multi-symplectic structure on the total exterior algebra bundle. Proc. R. Soc. Lond. A **462**, 1531–1551 (2006)
28. Bridges, T.J., Derks, G.: Unstable eigenvalues and the linearization about solitary waves and fronts with symmetry. Proc. R. Soc. Lond. A **455**, 2427–2469 (1999)
29. Bridges, T.J., Derks, G.: The symplectic Evans matrix, and the instability of solitary waves and fronts. Arch. Rational Mech. Anal. **156**, 1–87 (2001)
30. Bridges, T.J., Derks, G.: Linear instability of solitary wave solutions of the Kawahara equation and its generalisations. SIAM J. Math. Anal. **33**, 1356–1378 (2002)
31. Bridges, T.J., Derks, G.: Constructing the symplectic Evans matrix using maximally analytic individual vectors. Proc. R. Soc. Edinb. A **133**, 505–526 (2003)
32. Bridges, T.J., Reich, S.: Multi-symplectic spectral discretizations for the Zakharov-Kuznetsov and shallow water equations. Physica D **152–153**, 491–504 (2001)
33. Bridges, T.J., Hydon, P.E., Lawson, J.K.: Multisymplectic structures and the variational bicomplex. Math. Proc. Camb. Philos. Soc. **148**, 159–178 (2010)
34. Burchell, T.J.: Nonlinear wave equations, Clifford Analysis, and the stability of solitary waves. PhD Thesis, University of Surrey (2022)
35. Burchell, T.J., Bridges, T.J.: Symplectic transversality and the Pego-Weinstein theory. Adv. Math. **406**, 108524 (2022)
36. Buryak, A.V., Akhmediev, N.N.: Stability criterion for stationary bound states of solitons with radiationless oscillating tails. Phys. Rev. E **51**, 3572–3578 (1995)
37. Cai, J., Chen, J., Chen, M.: Efficient linearized local energy-preserving method for the Kadomtsev-Petviashvili equation. Discrete Contin. Dyn. Syst. Ser. B **27**, 2441 (2022)
38. Chardard, F.: Maslov index for solitary waves obtained as a limit of the Maslov index for periodic waves. C.R. Acad. Sci. Paris Ser. I **345**, 689–694 (2007)
39. Chardard, F.: Stabilité des Ondes Solitaires. PhD Thesis, École Normale Supérieure de Cachan, France (2009)
40. Chardard, F., Bridges, T.J.: Transversality of homoclinic orbits, the Maslov index and the symplectic Evans function. Nonlinearity **28**, 77–102 (2015)
41. Chardard, F., Dias, F., Bridges, T.J.: On the Maslov index of multi-pulse homoclinic orbits. Proc. R. Soc. Lond. A **465**, 2897–2910 (2009)
42. Chardard, F., Dias, F., Bridges, T.J.: Computing the Maslov index of solitary waves, Part 1: Hamiltonian systems on a four-dimensional phase space. Physica D **238**, 1841–1867 (2009)
43. Chen, C.-N., Hu, X.: Maslov index for homoclinic orbits of Hamiltonian systems. Ann. Inst. H. Poincaré Anal. Non Lin. **24**, 589–603 (2007)

44. Chen, R.M., Jin, J.: Transverse instability of the CH-KP-I equation. Ann. Appl. Math. **37**, 337–362 (2021)
45. Chen, Y., Song, S., Zhu, H.: The multi-symplectic Fourier pseudospectral method for solving two-dimensional Hamiltonian PDEs. J. Comput. Appl. Math. **236**, 1354–1369 (2011)
46. Chen, R.M., Fan, L., Wang, X., Xu, R.: Spectral analysis of the b-KP equation under transverse perturbations. Math. Ann. **390**, 6315 (2024). s00208-024-02907-8
47. Chhay, M., Dutykh, D., Clamond, D.: On the multi-symplectic structure of the Serre-Green-Naghdi equations. J. Phys. A Math. Theor. **49**, 03LT01 (2016)
48. Cho, Y.: Stability of gravity-capillary solitary waves on shallow water based on the fifth-order Kadomtsev-Petviashvili equation. Phys. Rev. E **98**, 012213 (2018)
49. Clamond, D., Dutykh, D.: Practical use of variational principles for modeling water waves. Physica D **241**, 25–36 (2012)
50. Clamond, D., Dutykh, D.: Multi-symplectic structure of fully-nonlinear weakly-dispersive internal gravity waves. J. Phys. A Math. Theor. **49**, 31LT01 (2016)
51. Clamond, D., Dutykh, D., Mitsotakis, D.: Variational Approach to Water Waves Modelling, IAHR Water Monographs. International Association for Hydro-Environment Engineering and Research (IAHR), Spain (2024)
52. Coppel, W.A.: Stability and Asymptotic Behaviour of Differential Equations. D.C. Heath and Co, Boston (1965)
53. Cornwell, P., Jones, C.K.R.T., Kiers, C.: Bifurcation to instability through the lens of the Maslov index. J. Dyn. Differ. Equ. **36**, 237–148 (2024)
54. Cox, G., Jones, C.K.R.T., Latushkin, Y., Sukhayev, A.: Morse and Maslov indices for multidimensional Schrödinger operators with matrix-valued potentials. Trans. Amer. Math. Soc. **368**, 8145–8207 (2016)
55. Cox, G., Curran, M., Latushkin, Y., Marangell, R.: Hamiltonian spectral flows, the Maslov index, and the stability of standing waves in the nonlinear Schrödinger equation. SIAM J. Math. Anal. **55**, 4998–5050 (2023)
56. Cox, G., Latushkin, Y., Sukhtayev, A.: Fredholm determinants, Evans functions and Maslov indices for partial differential equations. Math. Ann. **389**, 1963–2036 (2024)
57. Curran, M.: Hamiltonian spectral theory and the Maslov index. PhD Thesis, School of Maths & Stats, The University of Sydney (2024)
58. Das, M., Latushkin, Y.: Derivatives of the Evans function and (modified) Fredholm determinants for first order systems. Math. Nachr. **284**, 1592–1638 (2011)
59. Dauxois, T., Joubaud, S., Odier, P., Venaille, A.: Instabilities of internal gravity wave beams. Ann. Rev. Fluid Mech. **50**, 131–156 (2018)
60. de Gosson, M.: Symplectic Geometry and Quantum Mechanics. Birkhäuser, Basel (2006)
61. Deconinck, B., Pelinovsky, D.E., Carter, J.D.: Transverse instabilities of deep-water solitary waves. Proc. R. Soc. Lond. A **462**, 2039–2061 (2006)
62. Dirac equation, https://en.wikipedia.org/wiki/Dirac_equation
63. Drazin, P.G., Reid, W.H.: Hydrodynamic Stability. Cambridge University Press, Cambridge (2004)
64. Durń, A., Dutykh, D., Mitsotakis, D.: On the multi-symplectic structure of Boussinesq-type systems. I: Derivation and mathematical properties. Physica D **388**, 10–21 (2019)
65. Durń, A., Dutykh, D., Mitsotakis, D.: On the multi-symplectic structure of Boussinesq-type systems. II: Geometric discretization. Physica D **397**, 1–16 (2019)
66. Dutykh, D., Chhay, M., Clamond, D.: Numerical study of the generalised Klein-Gordon equations. Physica D **304**, 23–33 (2015)
67. Gaivao, J.P., Gelfreich, V.: Splitting of separatrices for the Hamiltonian-Hopf bifurcation with the Swift-Hohenberg equation as an example. Nonlinearity **24**, 677–698 (2011)
68. Gelfreich, V.G., Lazutkin, V.F.: Splitting of separatrices: perturbation theory and exponential smallness. Russ. Math. Surv. **56**, 499–558 (2001)
69. Gesztesy, F., Latushkin, Y., Makarov, K.A.: Evans functions, Jost functions, and Fredholm determinants. Arch. Rational Mech. Anal. **186**, 361–421 (2007)

70. Geyer, A., Liu, Y., Pelinovsky, D.E.: On the transverse stability of smooth solitary waves in a two-dimensional Camassa-Holm equation. J. Math. Pure Appl. **188**, 1–25 (2024)
71. Godey, C.: A simple criterion for transverse linear instability of nonlinear waves. C. R. Math. **354**, 175–179 (2016)
72. Grillakis, M., Shatah, J., Strauss, W.: Stability theory of solitary waves in the presence of symmetry, I. J. Funct. Anal. **74**, 160–197 (1987)
73. Grillakis, M., Shatah, J., Strauss, W.: Stability theory of solitary waves in the presence of symmetry, II. J. Funct. Anal. **94**, 308–348 (1990)
74. Groves, M.D., Haragus, M., Sun, S.M.: Transverse instability of gravity-capillary line solitary water waves. C. R. Ser. I **333**, 421–426 (2001)
75. Haragus, M., Wahlén, E.: Transverse instability of periodic and generalized solitary waves for a fifth-order KP model. J. Differ. Equ. **262**, 3235–3249 (2017)
76. Her, H.-L., Zhong, Q.: On Maslov-type index for general paths of symplectic matrices, `arXiv:2407.08433` (2024)
77. Howard, P.: The Maslov index and spectral counts for linear Hamiltonian systems on $\mathbb{R}$. J. Dyn. Differ. Equ. **35**, 1947–1991 (2021)
78. Hu, W., Deng, Z., Qin, Y.: Multi-symplectic method to simulate soliton resonance of $(2+1)-$dimensional Boussinesq equation. J. Geom. Mech. **5**, 295–318 (2013)
79. Infeld, E., Rowlands, G.: Nonlinear Waves, Solitons, and Chaos. Cambridge University Press, Cambridge (1990)
80. Infeld, E., Rowlands, G., Hen, M.: Three dimensional stability of Korteweg de Vries waves and solitons. Acta Phys. Polon. **A54**, 131–139 (1978)
81. Iwasaki, H., Toh, S., Kawahara, T.: Cylindrical quasi-solitons of the Zakharov-Kuznetsov equation. Physica D **43**, 293–303 (1990)
82. Janssen, P., Rasmussen, J.: Nonlinear evolution of the transverse instability of plane envelope solitons. Phys. Fluids **26**, 1279–1287 (1983)
83. Jiang, B., Wang, Y., Cai, J.: A new multisymplectic scheme for generalized Kadomtsev-Petviashvili equation. J. Math. Phys. **47**, 083503 (2006)
84. Jones, C.K.R.T.: Instability of standing waves for non-linear Schrödinger-type equations. Ergodic Theory Dyn. Syst. **8**, 119–138 (1988)
85. Kadomtsev, B.B., Petviashvili, V.I.: On the stability of solitary waves in weakly dispersive media. Sov. Phys. Dokl. **15**, 539–541 (1970)
86. Kapitula, T., Promislow, K.: Spectral and Dynamical Stability of Nonlinear Waves, Appl. Math. Sci., vol. 185. Springer, New York (2013)
87. Kataoka, T.: Transverse instability of interfacial solitary waves. J. Fluid Mech. **611**, 255–282 (2008)
88. Kataoka, T.: Transverse instability of surface solitary waves. Part 2. Numerical linear stability analysis. J. Fluid Mech. **657**, 126–170 (2010)
89. Kataoka, T., Tsutahara, M.: Instability of solitary wave solutions to long-wavelength transverse perturbations in the generalized Kadomtsev-Petviashvili equation with negative dispersion. Phys. Rev. E **70**, 016604 (2004)
90. Kataoka, T., Tsutahara, M.: Transverse instability of surface solitary waves. J. Fluid Mech. **512**, 211–221 (2004)
91. Kataoka, T., Tsutahara, M., Negoro, Y.: Transverse instability of solitary waves in the generalized Kadomtsev-Petviashvili equation. Phys. Rev. Lett. **84**, 3065–3068 (2000)
92. Kharif, C., Abid, M., Chen, Y.Y., Hsu, H.C.: A nonlinear Schrödinger equation for capillary waves on arbitrary depth with constant vorticity. HAL-04535263 (2024)
93. Kim, B.: Long-wave transverse instability of weakly nonlinear gravity-capillary solitary waves. J. Eng. Math. **74**, 19–28 (2012)
94. Kim, B., Akylas, T.R.: Transverse instability of gravity–capillary solitary waves. J. Eng. Math. **58**, 167–175 (2007)
95. Kivshar, Y.S., Pelinovsky, D.E.: Self-focusing and transverse instabilities of solitary waves. Phys. Rep. **331**, 117–195 (2000)

96. Klein, C., Saut, J.-C., Stoilov, N.: Numerical study of the transverse stability of line solitons of the Zakharov-Kuznetsov equations. Physica D **448**, 133722 (2023)
97. Krechetnikov, R.: Transverse instability of concentric water waves. J. Nonlinear Sci. **34**, 66 (2024)
98. Kuznetsov, E.A., Rubenchik, A.M., Zakharov, V.E.: Soliton stability in plasmas and hydrodynamics. Phys. Rep. **142**, 103–165 (1986)
99. Latushkin, Y., Sukhtaiev, S.: The Maslov index and the spectra of second order elliptic operators. Adv. Maths. **329**, 422–486 (2018)
100. Levandosky, S.P.: A stability analysis of fifth-order water wave models. Physica D **125**, 222–240 (1999)
101. Levandosky, S.P.: Stability of solitary waves of a fifth-order water wave model. Physica D **227**, 162–172 (2007)
102. Li, H., Sun, J.: A new multi-symplectic scheme for the generalized Kadomtsev–Petviashvili equation. Comput. Phys. Comm. **183**, 1956–1965 (2012)
103. Li, H., Wang, Z.: Dynamics of interfacial gravity-capillary waves in three-dimensional fluids of great depth. Comput. Fluids **231**, 105151 (2021)
104. Liu, Y., Qin, M.: Multisymplectic geometry and multisymplectic Preissman scheme for the KP equation. J. Math. Phys. **43**, 4060–4077 (2002)
105. Long, Y.: Index Theory for Symplectic Paths with Applications. Birkhäuser, Basel (2002)
106. Longuet-Higgins, M.S., Tanaka, M.: On the crest instabilities of steep surface waves. J. Fluid Mech. **336**, 51–68 (1997)
107. Lounesto, P.: Clifford Algebras and Spinors. LMS Lecture Notes, vol. 239. Cambridge University Press, Cambridge (1997)
108. Maddocks, J.H., Sachs, R.L.: Constrained variational principles and stability in Hamiltonian systems. In: IMA Volumes in Mathematics and its Applications, vol. 63, pp. 231–264. Springer, New York (1995)
109. Marsden, J.E., Ratiu, T.: Introduction to Mechanics and Symmetry, Texts in Appl. Maths., vol. 17. Springer, Berlin (1994)
110. Morita, S.: Geometry of Differential Forms. AMS, Providence (2001)
111. Munro, S., Parkes, E.J.: The derivation of a modified Zakharov-Kuznetsov equation and the stability of its solutions. J. Plasma Phys. **62**, 305–317 (1999)
112. Murashige, S., Choi, W.: Linear stability of transversely modulated finite-amplitude capillary waves on deep water. Stud. Appl. Math. **146**, 498–525 (2021)
113. Ostrovskii, L.A., Shrira, V.I.: Instability and self-refraction of solitons. J. Exp. Theoret. Phys. **44**, 738–743 (1976)
114. Paddick, M.: Transverse nonlinear instability of Euler-Korteweg solitons. Ann. Fac. Sci. Toulouse Math. **26**, 23–48 (2017)
115. Park, B., Cho, Y.: Two-dimensional gravity-capillary solitary waves on deep water: generation and transverse instability. J. Fluid Mech. **834**, 92–124 (2018)
116. Pego, R.L., Sun, S.M.: On the transverse linear instability of solitary water waves with large surface tension. Proc. R. Soc. Edinb. **134**, 733–752 (2004)
117. Pego, R.L., Weinstein, M.I.: Eigenvalues, and instabilities of solitary waves. Philos. Trans. R. Soc. Lond. A **340**, 47–94 (1992)
118. Pelinovsky, D.E.: A mysterious threshold for transverse instability of deep-water solitons. Math. Comput. Simul. **55**, 585–594 (2001)
119. Pelinovksy, D.E.: Normal form for transverse instability of the line soliton with a nearly critical speed of propagation. Math. Model. Nat. Phenom. **13**, 23–43 (2018)
120. Pelinovsky, D.E., Shimabukuro, Y.: Transverse instability of line solitary waves in massive Dirac equations. J. Nonlinear Sci. **26**, 365–403 (2016)
121. Porteous, I.: Clifford Algebras and Classical Groups. Cambridge University Press, Cambridge (1995)
122. Pushkar, P.E.: Maslov index and symplectic Sturm theorems. Func. Anal. Appl. **32**, 172–182 (1988)

123. Ratliff, D.J.: Flux singularities in multiphase wavetrains and the Kadomtsev-Petviashvili equation with applications to stratified hydrodynamics. Stud. Appl. Math. **142**, 109–138 (2019)
124. Ratliff, D.J., Bridges, T.J.: Phase dynamics of periodic waves leading to the Kadomtsev-Petviashvili equation in 3+1 dimensions. Proc. R. Soc. Lond. A **471**, 20150137 (2015)
125. Rousset, F., Sun, C.: Transverse linear stability of one-dimensional solitary gravity water waves. arXiv:2402.11115 (2024)
126. Rousset, F., Tzvetkov, N.: Transverse nonlinear instability of solitary waves for some Hamiltonian PDE's. J. Math. Pures Appl. **90**, 550–590 (2008)
127. Rousset, F., Tzvetkov, N.: Transverse nonlinear instability for two-dimensional dispersive models. Ann. I. H. Poincaré AN **26**, 477–491 (2009)
128. Rousset, F., Tzvetkov, N.: A simple criterion of transverse linear instability for solitary waves. Math. Res. Lett. **17**, 157–169 (2010)
129. Rousset, F., Tzvetkov, N.: Transverse instability of the line solitary water-wave. Invent. Math. **184**, 257–388 (2011)
130. Rousset, F., Tzvetkov, N.: Stability and instability of the KDV solitary wave under the KP–I flow. Commun. Math. Phys. **313**, 155–173 (2012)
131. Rypdal, K., Rasmussen, J.J.: Stability of solitary structures in the nonlinear Schrödinger equation. Phys. Scr. **40**, 192 (1989)
132. Sipcic, R., Benney, D.J.: Lump interactions and collapse in the modified Zakharov-Kuznetsov equation. Stud. Appl. Math. **105**, 385–403 (2000)
133. Souriau, J.-M.: Construction explicite de l'indice de Maslov, Applications, Group Theoretical Methods in Physics (4th Inter. Colloq. Nijmegen, 1975), Springer Lect. Notes in Physics, vol. 50, pp. 117–148. Springer, Berlin (1976)
134. Sprenger, P., Bridges, T.J., Shearer, M.: Traveling wave solutions of the Kawahara joining distinct periodic waves. J. Nonlinear Sci. **33**, 79–118 (2023)
135. Treschev, D.V.: Separatrix splitting from the point of view of symplectic geometry. Mat. Z. **61**, 890–906 (1997)
136. Varma, D., Mathur, M., Dauxois, T.: Instabilities in internal gravity waves. Math. Eng. **5**, 1–34 (2023)
137. Wang, Z., Li, C.: On modified b-KP systems and generalisations. Theor. Math. Phys. **209**, 1693–1716 (2021)
138. Wang, Y., Wang, B., Yang, H., Wang, Y.: An artificial boundary condition for the multisymplectic Preissman scheme. J. Phys. Soc. Japan **73**, 1457–1463 (2004)
139. Yamazaki, Y.: Stability of line standing waves near the bifurcation point for nonlinear Schrödinger equations. Kodai Math. J. **38**, 65–96 (2015)
140. Yamazaki, Y.: Stability for line solitary waves of Zakharov-Kuznetsov equation. J. Differ. Equ. **262**, 4336–4389 (2017)
141. Yamazaki, Y.: Center-stable manifolds around line solitary waves of the Zakharov-Kuznetsov equation. J. Dyn. Differ. Equ. **36**, 871–914 (2024)
142. Zakharov, V.E., Kuznetsov, E.A.: Three-dimensional solitons. Soviet Phys. JETP **39**, 285–286 (1974)
143. Zakharov, V.E., Rubenchik, A.M.: Instability of waveguides and solitons in nonlinear media. Sov. Phys. JETP **38**, 494–500 (1974)

Index

B
Boussinesq equation, 21, 18

C
Camassa–Holm equation, 18
Clifford algebra, 39

D
d'Alembertian, 13, 103
Dedonder–Weyl theory, 8
Dirac equation, 31
Dirac equations, PS-type (2.10), 19

E
Energy–momentum method, 5, 6, 55
Evans function, 11, 12, 20
  derivative, 87
  equivalence, 83
  exterior algebra form, 75
  invariance, 83
  matrix form, 80, 101
Example, 21, 103

F
Factor χ, 70

G
Gross–Neveu model, 20

H
Hamiltonian, 3, 5
Hypothesis H1, 53, 99
Hypothesis H2, 56, 99
Hypothesis H3, 64, 100
Hypothesis H4, 65, 100
Hypothesis H5, 70, 100
Hypothesis H6, 75, 101
Hypothesis H7, 85, 101

I
Instability, longitudinal, 17

K
Kawahara equation, 16
KD equation, 19
Ker(**L**), 64
Klein–Gordon equation, 49
Korteweg fluids, 22
KP equation, 1, 15, 137
  BBM form, 24
  b-form, 19
  fifth-order, 16, 133
  generalised, 18
  Lagrangian, 137
  multisymplectic, 47, 59
  rotation-modified, 19

L
Lagrangian Grassmannian, 149
Lagrangian subspace, 66, 68
Lazutkin–Treschev invariant, 12, 67, 100, 113

T. J. Burchell, T. J. Bridges, *Transverse Instability of Solitary Waves*, Mathematical Physics Studies, https://doi.org/10.1007/978-981-95-2046-6

M
Maslov index, 12, 26, 149, 153
Multisymplectic, 7
  canonical, 8, 31, 53, 99, 139
  complex form, 45
  Dirac form, 36
  KP, 139

N
NLS equation, 24
NLS equation, in, 21, 25

O
Orientation, 69

R
Reversibility, 56

S
Solitary wave, 1, 54
Souriau mapping, 149
Spectral ansatz, 3, 11, 16
Stability, longitudinal, 6
Stability, nonlinear, 23
Stability, transverse, 1
Stable subspace, E^s, 66
Symplectic, operator **J**, 53
System at infinity, 64
  adjoint, 65
  eigenvalues, 65
  eigenvectors, 65

T
TEA bundle, 36
TEA bundle, canonical form, 37
Transversality, 67

U
Unstable subspace, E^u, 66

W
Water wave equations, 27

Z
ZK equation, 1, 22

MIX
Papier aus verantwortungsvollen Quellen
Paper from responsible sources
FSC® C105338

If you have any concerns about our products,
you can contact us on
ProductSafety@springernature.com

In case Publisher is established outside the EU,
the EU authorized representative is:
Springer Nature Customer Service Center GmbH
Europaplatz 3, 69115 Heidelberg, Germany

Printed by Libri Plureos GmbH
in Hamburg, Germany